THE NAVAL ARMS TRADE

STRATEGIC ISSUE PAPERS

The SIPRI series *Strategic Issue Papers* focuses on topical issues of significance for the future of international peace and security. The naval arms race is emerging as one of the central security and arms control issues of our time. SIPRI has therefore chosen *The Naval Arms Trade* as the third in the series. The studies address problems relating to arms reduction, the spread of arms, military and political strategy and the impact of technology on the conduct of peaceful East–West relations. The books are concise, with short production times so as to make a timely input into current debates.

sipri
Stockholm International Peace Research Institute

SIPRI is an independent institute for research into problems of peace and conflict, especially those of arms control and disarmament. It was established in 1966 to commemorate Sweden's 150 years of unbroken peace.

The Institute is financed mainly by the Swedish Parliament. The staff, the Governing Board and the Scientific Council are international.

The Governing Board and the Scientific Council are not responsible for the views expressed in the publications of the Institute.

sipri
Stockholm International Peace Research Institute

Pipers väg 28, S-171 73 Solna, Sweden
Cable: PEACERESEARCH STOCKHOLM
Telephone: 46 8/55 97 00

The Naval Arms Trade

Ian Anthony

sipri

Stockholm International Peace Research Institute

OXFORD UNIVERSITY PRESS
1990

Oxford University Press, Walton Street, Oxford OX2 6DP
Oxford New York Toronto
Delhi Bombay Calcutta Madras Karachi
Petaling Jaya Singapore Hong Kong Tokyo
Nairobi Dar es Salaam Cape Town
Melbourne Auckland
and associated companies in
Berlin Ibadan

Oxford is a trade mark of Oxford University Press

Published in the United States
by Oxford University Press, New York

British Library Cataloguing in Publication Data
Anthony, Ian
The naval arms trade.—(Strategic issue papers)
1. Foreign trade in military equipment
I. Title   II. Stockholm International Peace Research Institute   III. Series
382'.456234
ISBN 0-19-829137-X

Library of Congress Cataloging in Publication Data
Anthony, Ian.
The naval arms trade/Ian Anthony.
(Strategic issue papers)
1. Navies.   2. Munitions.   3. Sea-power.   I. Title.   II. Series.
VA40.A48   1989   382'.4623825—dc20   89-23037
ISBN 0-19-829137-X

Typeset and originated by Stockholm International Peace Research Institute
Printed and bound in Great Britain by
Biddles Ltd., Guildford and King's Lynn

Contents

Part II. Case studies

Part III. Conclusions

Preface

The naval arms race is emerging as one of the central security issues of our time. Coastal states have continued to invest in the arms that they consider appropriate to their needs without joint consideration of the consequences of the proliferation of naval technologies. Given the economic importance of freedom of navigation to all trading nations and the history of navies as instruments of power projection, the spread of naval armaments has increased the potential level of violence inherent in the conduct of 'gunboat diplomacy'. As the economic importance of the resources in and under the seas has grown, a host of new areas of potential conflict between neighbours have emerged. However, since 1945 naval armaments have been kept out of arms control negotiations.

In seeking a workable global regime for enhanced security at sea, discussion of naval arms control has largely concerned the navies of the United States, the Soviet Union, France and the UK. This book explains the imperatives and national interests of other coastal states, thus addressing a dimension of international maritime security which is of growing consequence to the arms control debate. The purpose is descriptive rather than prescriptive. The focus of this book is how the transfer of naval technologies between countries has changed and in turn been changed by the global maritime environment. Its function is to describe which countries are participants in the naval arms trade, to explain why they buy what they buy and to examine some of the consequences of their actions.

SIPRI
June 1989

Dr Walther Stützle
Director

Acknowledgements

The author would like to acknowledge the assistance of several people in the preparation of this book. Gillian Stanbridge edited the manuscript and set the entire monograph in camera-ready format. Early drafts of the manuscript benefited enormously from the comments of Rear Admiral Richard Hill and Dr Michael Brzoska, although full responsibility for the final product lies with the author.

Acronyms

AALS	Amphibious assault landing ship
AMEC	Australian Marine Engineering Corporation
AMPS	Autonomous Marine Power System
ANZUS	Australia–New Zealand–United States (alliance)
ASM	Air-to-surface missile
ASROC	Anti-submarine rocket
ASW	Anti-submarine warfare
CER	Closer Economic Relations (agreement)
CFE	(Negotiation on) Conventional (Armed) Forces in Europe
CIWS	Close-in weapon system
EEZ	Exclusive Economic Zone
EMPAR	European Multifunction Phased Array Radar
FAC	Fast attack craft
FAMS	Family of Missile Systems
FPDA	Five Power Defence Arrangement
HDW	Howaldtswerke Deutsche Werft
IAF	Indian Air Force
IEPG	Independent European Program Group
LAMS	Local Area Missile System
LOS	Law of the Sea
MBFR	Mutual and Balanced Force Reductions
MEKO	Mehrzweck Kontainer
MOU	Memorandum of Understanding
MR	Maritime reconnaissance
MSC	Minesweeper (coastal)
MSO	Minesweeper (offshore)
NAAWS	NATO Anti-Air Warfare System
NATO	North Atlantic Treaty Organization
NFR	NATO Frigate Replacement
NIAG	NATO Industrial Advisory Group

NICs	Newly industrializing countries
NNAG	NATO Naval Armaments Group
OPV	Offshore patrol vessel
PMO	Project Management Office
RAAF	Royal Australian Air Force
RAM	Rolling Airframe Missile
RAN	Royal Australian Navy
RFP	Request for Proposals
RRF	Ready Reaction Force
SAARC	South Asian Association for Regional Cooperation
SAM	Surface-to-air missile
ShShM	*Sh*ip-to-*sh*ip *m*issile
SMART	Signaal Multibeam Acquisition Radar for Targeting
SRAAWS	Short Range Anti-Air Warfare System
SSN	Nuclear-powered attack submarine
SSTDS	Surface Ship Torpedo Defence System
TNSW	Thyssen Nordseewerke
TRUMP	Tribal Class Update and Modernization Program
VLF	Very low frequency
VSEL	Vickers Shipbuilding and Engineering Consortium

Part I. Overview

1. Introduction

After assuming office as Second Sea Lord, it did not take me long to find
out that Mr Churchill, the First Sea Lord, was very apt to express strong
opinions upon purely technical matters . . . [H]is fatal error was his inability
to realize his own limitations as a civilian quite ignorant of naval affairs.[1]
(Admiral of the Fleet Earl Jellicoe of Scapa)

I. Why study the naval arms trade?

While within the naval powers themselves a relatively small community of interested parties has always concerned itself with issues of
maritime security, neither this group nor the literature that it has generated has entered the mainstream of the strategic studies debate.
Naval policy and naval affairs have been seen as areas of national
policy, largely unaffected by multilateral actions and offering little
scope even for discussion in an international context. The lack of
interest in superpower naval arms control has been mirrored by a
relative neglect of the issue of naval arms sales, their implications and
whether their control is possible or even desirable. Within the literature on the arms trade, there has been remarkably little discussion of
the transfer of naval weapon systems.[2]

The low level of interest in the naval arms trade is in part explicable
by the fact that interest in arms transfer issues has been strongest
where sales concern either the Soviet Union or the United States. As
indicated in table 1.1, naval systems represent a relatively small percentage of the total arms sales by the superpowers. For the major
European arms exporters, naval systems occupy a much more central
role in the balance of their sales.

The trade in naval systems accounts for as large a percentage of
total transfers as does the trade in systems for use on land or in the air.
Naval forces are an essential component of national security in all
coastal states and given that this is so, the importance of the naval
arms trade becomes even clearer when one notes that—with the
exception of the two superpowers, France and the UK—every country

Table 1.1. Naval exports as a percentage of their total arms exports 1979-88[a]

Supplier	USA	France	USSR	FRG	UK	Italy
Percentage	24	57	9	51	45	78

[a] The definition of naval equipment applied here is explained more fully in chapter 2. However, it basically includes ships of 100 tonnes displacement or more, fixed- and rotary-wing maritime aircraft, naval guided missiles and their associated fire control systems.

depends to some extent on imported technology to equip naval forces. Even this characterization may be insufficiently sweeping. In May 1988, the National Steel and Shipbuilding Company (NASCO) of San Diego signed an agreement with Fincantieri Cantieri Navali of Italy covering the joint production and marketing of Italian submarine, frigate and corvette designs. However, any designs adopted must be built by companies with 51 per cent US ownership.[3] In 1988, the United States also placed an order with Italy for the construction of an enlarged version of the Lerici class minehunter.[4] The Soviet Union has also imported a significant number of warships from Polish shipyards, in particular the Polnochny class amphibious landing craft. Poland and, to a lesser extent, the German Democratic Republic (GDR), also construct significant numbers of auxiliaries for the Soviet Navy.[5] In addition, all navies import certain sub-systems and components, the loss of access to which would be both disruptive and expensive. While it might be unfair to say that France and the UK *depend* on imported sub-systems, these countries have not been immune from the growing inter-dependence of NATO and intra-European defence industrial production. As one commentator has observed:

Technology has its own logic which is in the process of subverting the old model of the purely national, arsenal-type defence industry. The injection of market mechanisms, not only in competition for export markets but also on our home ground, has become a necessity and a possibility. Despite the political and strategic constraints which weigh on the European countries . . . the notion of a European defence industry market is no longer a fantasy.[6]

The naval arms trade is important even to major naval powers, but it is the central or exclusive element of naval procurement in most countries. Many countries that appear to satisfy a significant proportion of their equipment needs from domestic production change the nature of their participation in the naval arms trade rather than withdrawing from it. Third World countries such as Brazil, India and South Korea have invested heavily in naval shipbuilding but none has been able to develop indigenously such basic and essential sub-systems as engines, radars and communications systems.

In a case of emergency, it is probable that a wide range of purely air force and army systems could play a role in maritime operations, as could civilian assets such as merchant fleets. Equally, systems operated by the navy would play a part in coastal air defence and might also have a role in operations on land. For this reason, a rigid division of the arms trade into sea, land and air systems is artificial, and therefore in this monograph consideration is also given to the transfer of other systems such as aircraft, missiles and coastal defence systems devoted to maritime operations. Defined in this broad manner, it is self-evident that the naval arms trade is an important part of the overall arms market.

For the purposes of this monograph, the definition of the naval arms trade encompasses not only transfers of naval vessels and aircraft, and the trade in their associated weapon systems and sensors, but also programmes of collaborative naval production. This broad definition includes a vast number of transactions, many of them involving the transfer of civilian technologies with military applications. In addition, it includes equipment used in joint navy–air force–army–coast guard operations (and the effectiveness of command and control in those joint operations), merchant fleets and the civil maritime infrastructure.

There are in fact only two delimiting criteria attached to the arms trade here. First, the arms trade as defined requires that technology is transferred from country to country. Excluded from consideration are purely national military programmes where research, design, development and production are carried out within one country. This is still trade in the sense that there is a customer (the national armed forces) and a retailer or producer. However, transactions within one country respond to different sets of criteria from those involving the international trade. This kind of procurement does not involve sensitive decisions about the impact of arms deals on foreign relations; neither are the possible negative consequences of increasing the military capabilities of another country relevant. Domestic relationships between ministries of defence and national industry are certainly politicized, but the issues that they involve are those of the political economy—budget planning, the role of the state in the economy, employment, regional economic development, cost effectiveness and so on.

Second, transfers must be under the control of the government in both the supplier and recipient. countries Transactions which involve private groups or individuals, those which are made to opposition forces within any country and those which take place without government consent are excluded.

If it is necessary to include new forms of naval procurement and a very broad range of equipment in the assessment of the naval arms trade, it is equally important not to ignore more traditional transactions. The international market for military equipment is not one that changes rapidly, and therefore, it is appropriate to begin with an overview of the arms market as it looks now. Part I of this book is devoted to analysis of the global naval arms market and some of the more important trends within it. In chapter 2, this overview is laid out according to six different categories of supplier and recipient. These are: sellers of major, new naval systems; sellers of major, second-hand naval systems; sellers of naval sub-systems; buyers of major, new naval systems; buyers of major, second-hand naval systems; and buyers of naval sub-systems. These categories are not mutually exclusive, and it would be possible to belong to all six. In chapter 2 each of these different types of actor is examined in turn to give a better picture of the overall structure of the naval arms trade.

Identifying the main actors involved in the international flow of naval systems points to some of the underlying dynamics of the trade, but it does little to explain them. Understanding changes in the global balance brought about by technology transfer cannot involve disentangling transfers from a wider context. Participation in the naval arms trade is closely related to wider issues of naval power. Not only have new states developed navies, but some have sought alternative means of performing naval missions historically performed by surface vessels—notably through the purchase of submarines and maritime aircraft. The major suppliers of these systems are traditional naval powers and countries dependent on maritime trade, for which the ability to use the seas without restriction has been seen as an important national interest. Therefore, at this general level the naval arms trade is linked to the issues of establishing effective defence forces, the use of the sea as a maritime resource or for access to non-adjacent areas and the ability of naval powers to project power within and beyond their immediate environment.

Clearly several political and technological developments are occurring simultaneously.[7] Chapter 3 is devoted to studies of particular trends in naval development which have been reflected in transfers of specific types of system. These trends are the sale of maritime aircraft, submarines and equipment used for low-level maritime operations and maritime constabulary duties.

The first of these trends reflects the increasing range of weapons that can be brought to bear in naval operations. In a growing number of countries, issues of maritime security are not the exclusive preserve of the navy. Events such as the loss of six British ships, including four major surface combatants, during the Falklands/Malvinas War of 1982 and the attack on the frigate *USS Stark* in the Persian Gulf on 17 May 1987 have highlighted the importance of air forces and air-launched guided missiles to maritime security. These incidents underline the rapid changes in the nature of naval weapon technologies that have taken place during the 1970s and 1980s. Not only have the systems already in service increased their range, payload, accuracy and reliability, but a growing number of new weapon systems are now deployed by the navies of the world.

The second major trend examined in chapter 3 is the change in the political framework of maritime affairs. In the 1970s and 1980s an increasing number of states have begun to identify, define and exercise new maritime responsibilities in response to the United Nations Convention on the Law of the Sea signed in Jamaica in December 1982.[8] Part V of the United Nations Convention on the Law of the Sea (LOS) defined an exclusive economic zone (EEZ), a sea space out to a limit of 200 nautical miles (370 km), measured from the baseline of coastal seas. Some countries have assumed limited rights to exploit resources in this zone and noted that the Convention places on them a responsibility to exercise constabulary duties in this sea space.[9] Changing attitudes have also created new choices in resource allocation for coastal states. More traditional naval tasks such as coastal defence still have to be fulfilled, but added to these are other maritime responsibilities, such as the surveillance and policing of EEZs and newer and different security challenges, such as the elimination or control of drug trafficking, piracy and terrorism.

These changes in both the technology and politics of the maritime environment are taking place outside any effective multinational control. The Law of the Sea Convention has been ratified by too few states for it to enter into force, and adherence to its principles is purely voluntary. As of 2 February 1989 37 countries have ratified the Law of the Sea Convention. Sixty ratifications are needed for it to enter into force.[10]

II. The naval arms trade and arms control

Naval arms control is not a central issue in this monograph, reflecting the hostility to the concept of limiting naval forces among major powers, in general, and the United States, in particular. Even in the changed international environment that has emerged since the election of General Secretary Mikhail Gorbachev, one suspects that advocates of naval arms control are trying to push a large rock up a steep hill. However, although in the past international efforts to control naval armaments have been frustrated, developments in the naval arms trade underline a growing need to draw conventional military technology transfers into the arms control process. Moreover, although naval programmes are difficult to cancel once initiated, this is not impossible. As a consequence of the 1922 Washington Treaty the United States scrapped and destroyed battleships which were built and in the final process of fitting out. In the 1950s the Soviet Union cancelled a series of programmes.[11]

The issues which are examined here will be among those that determine whether a naval arms control process will begin or not and how that process will unfold. A close relationship exists between the development of naval forces and foreign policy in major powers, in particular among those of the Western alliance. Historically the importance of being able to conduct successful naval operations has outweighed the incentives for arms control. As one observer has put it, 'the facts of economic and technological development . . . set up an insurmountable barrier to the imposition of effective restraints on war at sea, and [re-emphasized] in the minds of European statesmen and public opinion the permanence of the organic relationship between maritime trade and naval war'.[12]

The importance of the role of sea power in global affairs has been much debated. One school has gone so far as to suggest a correlation between superior naval power and world hegemony.[13] Maritime strategists such as Corbett in the UK and Mahan in the United States were equally clear that the ability to control maritime communications and destroy commercial activity in enemy countries was a central feature of first British and subsequently US power. This organic relationship may be changing in ways which make naval arms control less unpalatable for maritime powers.

The issues that are relevant here include the following. First, major powers may in future find it difficult to sustain their naval forces at current levels, and this book examines changes in the way that naval procurement is taking place—in particular, the growing need for collaboration where projects are too expensive or technologically difficult for countries to undertake alone. Second, the growth of new naval powers and the spread of new naval weapons may circumscribe the naval activities of traditional naval powers. Third, historically stable supplier–ecipient relationships may be challenged by the emergence of new suppliers whose motivations appear primarily commercial, rather than responding to political criteria.

It is clear from this that the arms trade touches on a variety of political, economic and strategic issues, and disaggregation has been a traditional approach in the literature in order to make such a wide subject manageable. There have been regional arms trade studies, studies of arms trade to the Third World as opposed to that within the major alliances or between industrialized countries and studies of the trade in specific weapon systems or capabilities—in particular in the area of aerospace. The naval arms trade cannot be divided up in quite the same way. Criteria such as geographical separation, on which regional studies depend, are difficult to apply given the unique characteristics of sea power as an instrument of extra-regional force projection by major powers. The fact that ships must be allowed freedom of movement around the globe has been a touchstone of naval policy among major naval powers.

Discussing the trade in terms of specific systems or technologies is also made more difficult by the particular nature of major naval platforms. One author suggests that the acquisition of large naval vessels should not be viewed in the same way as the procurement of aircraft or tanks, but that a better unit of comparison would be an air base or an army camp.[14] Perhaps this is stretching the point, but it is the case that platforms weighing thousands of tonnes can support a considerable and diverse host of smaller platforms (including fixed-wing aircraft and helicopters), different weapon systems, sensors and communications. These different systems are integrated with one another on the same platform or on a small number of platforms in a task force, making major naval platforms self-contained in a sense that no land- or air-based system duplicates.

In dealing with these unique characteristics, most past work on the naval arms trade has focused on one of three characteristics: first, on changes in naval weapon system design and technologies; second, on the economic and industrial imperatives of the principal naval producers;[15] and third, on the diffusion of naval power to developing countries and its implications.[16] These three schools have concerned themselves with different elements of the naval arms trade. The first is principally a military-technological approach. The second is also technical in its orientation, but primarily concerns itself with the industrial and economic imperative behind the naval procurement policies of major powers. The third is essentially concerned with the political utility of maritime power.

All of these elements combine in shaping the arms trade. Therefore, in part II of the book, a series of case studies is used to illustrate the linkages between broad groups of issues in a specific context. These issues are of two kinds: those which are primarily political in their nature and those which are primarily economic and technological.

III. Political features of the naval arms trade

Historically, naval forces have not only been acquired for coastal defence. They have also been seen as an effective tool with which to support an offensive strategy and also as a political instrument. The application of naval force has been used in pursuit of economic and political advantage. The arms trade is interesting partly because it appears to reduce the ability of major naval powers to use naval force effectively. It is important to underline that discussing any aspect of the arms trade means concentrating on one element of a much larger picture. This is no less true for naval systems, and tracking the movement and location of weapon systems has great limitations as a method of assessing sea power.

Discussions of 'hardware' can make an important contribution where they make military and technical issues accessible to a civilian and non-specialist audience. However, a technological or system-oriented approach is less helpful where the implications of military programmes rest primarily on perceptions of political relationships. It is difficult to disagree with the assessment of Rear Admiral John Richard Hill that since power is a diverse, sometimes ill-defined unquantifiable thing, 'the search for classes or categories of power is most unlikely to end in a statistical table'.[17] Sea power is in fact the product of three elements—naval forces, merchant shipping and bases.[18]

The study of the arms trade generally only makes sense where it is linked to a specific context. Nevertheless, studies of naval inventories have, on occasion, been used to evaluate military capabilities with the intention of deducing the relative power of naval forces. One of the manifestations of this is a tendency towards the construction of 'hierarchies' of naval power and the discussion of the extent to which the diffusion of particular naval weapon systems—especially anti-ship missiles—has altered the naval balance.[19] The classification of navies according to their capabilities has appealed to a diverse group of naval analysts linked by their tendency to focus on the applications of naval forces, rather than the constituent parts of the forces themselves.[20] This can be a sophisticated and complicated exercise, incorporating a wide range of factors, and has been attempted on occasion, notably by European authors trying to grapple with defining the operational and political options realistically open to a 'medium' power. However, the exercise has rarely led to a satisfactory conclusion, reflecting a consensus among this group that it is impossible to measure sea power with purely quantitative indicators such as tonnage, equipment inventories, degree of technological sophistication, fire-power or expenditure.

The same exercise of 'measuring' relative naval power has been undertaken recently by a different group of authors arguing that the ability of major navies to project naval power has been threatened by the growing power of relatively small navies and that medium-sized navies, such as those of China and India, have altered the overall balance of naval forces.[21]

Predicting the power of a country by evaluating equipment inventories is a dangerous enterprise, however sophisticated the approach. The difficulty is underlined by the fact that there are so many historical occasions where predicted outcomes simply proved to be wrong. Clearly superior forces sometimes fail to achieve their objectives. Just a few examples from recent British experience run as follows: in 1958 and between 1973 and 1976, efforts by the Royal Navy to coerce Icelandic fishermen into accepting the British interpretations of regulations concerning fishing failed. Between 1966 and 1968 the Royal Navy tried with mixed success to turn away tankers carrying oil to

Rhodesia. In 1974, the Turkish seaborne invasion of Cyprus succeeded in the face of British commitments to the defence of the existing settlement. In 1978, British, French and US forces held joint exercises off the Horn of Africa during Ethiopia's war with Somalia without any impact on the conflict. On the other hand, operations which look very uncertain of success sometimes work, the recapture by the UK of the Falklands/Malvinas islands in 1982 being the clearest example.

The purpose of ranking naval powers is to give an indication of the current military balance at sea and to allow changes in the nature of the global balance to be described. Over time there have been significant changes in the balance of forces at sea, notably the increased capabilities of the Soviet Navy and naval air forces and latterly the acquisition of a submarine-based strategic nuclear force by the People's Republic of China. The global hierarchy of naval power has changed significantly since 1945 only as a consequence of naval programmes that have unfolded over 15–20 years or more. During the 1970s the Soviet Union emerged as a major naval power as a consequence of programmes initiated in the mid-1960s. Moreover, an expansion of naval capabilities pre-supposes the existence of a shore-based infrastructure that requires decades to develop. In both India and China the expansion of naval capabilities is currently the focus for much debate. However, in each case current equipment programmes have followed the expansion of shore-based facilities during the 1970s.

Quantitative indicators in combination form one component of a complete maritime picture, the understanding of which has to include two other groups of factors: first, the specific context—itself a function of geography, regional political relationships and the regional policies of major powers; and second, the location of naval issues within a wider national security picture. The United Nations study of the naval arms race summed these components up succinctly when it reported that

Naval forces do not exist independently of others, and must be considered in an overall military context. Naval parity or balance in itself means nothing. Hence naval disarmament has to be seen in a broader context. Secondly, this fact, and the geographical situation peculiar to individual States might lead a State to take unilateral action as regards limitation of naval forces and armaments so as to preserve an overall military balance, and hence to disregard numerical equality.[22]

The classification of countries according to their different stages of development has been a traditional approach to the question of context. The usual categories include the separation of the Third World from the industrialized world or the separation of neutral and non-aligned countries from members of the major alliances. In the field of military equipment or technology transfers, relationships remain closest between countries with a rough identity of views, often reflected in a treaty. For this reason, arms transfer relationships are interpreted by third parties as unambiguous indicators of a close bilateral relationship. Moreover, shifts within an arms transfer relationship are used to measure whether that overall bilateral relationship is improving or deteriorating and carry a weight which is seldom attached to other forms of transaction. For all these reasons, procurement decisions are used by governments to convey messages both to each other and to third parties about the nature of bilateral relationships.

There are important differences between those countries which are allied and those which are not. Moreover, the members of the major alliances include the largest naval shipbuilders and the main exporters of naval equipment. At the same time the trade in finished naval systems has been principally a North–South trade. The market for naval vessels in the major shipbuilding countries has been closed to foreign suppliers while as late as 1986 one analyst was able to write 'if there is an international market for warships, we must obviously look for it outside the two major alliances and China . . . The outside market comprises the navies of the Third World and those of non-aligned industrial states'.[23]

This concentration on industrial countries as sellers and the Third World as buyers fails to reflect the complexity of the naval arms trade. The navies of many industrial countries were rebuilt after 1945 by importing ships from the Soviet Union, the United Kingdom and, in particular, the United States. On the other hand, since so few countries have fully autonomous shipbuilding industries, there is also an enormous trade in naval sub-systems and technology incorporated in ships, which is both economically valuable and important for its political and military implications. In this field the major naval powers have dominated the market as recipients.

IV. The naval arms trade and the weapon acquisition process

Looking on arms transfers as isolated transactions is also misleading because the most obvious characteristic of the naval arms trade is the sheer complexity of the process of naval procurement. Programmes unfold over an extended time-scale—frequently closer to 15 years than 10. This fact makes the planning of the naval procurement process mandatory and also introduces a degree of flexibility into the process. Programmes must respond to national budget plans and may be vulnerable to being extended over a longer and longer period. Moreover, the cost and complexity of naval systems create pressures to co-operate in the development and production of naval systems.

Decisions on naval force planning reflect the concerns of the political economy in both buyer and seller. These transactions involve the expenditure of large quantities of national resources, although not necessarily hard currency. Profit maximization, cost-effectiveness and macro-economic management clearly have a bearing on the outcome of programme plans. However, in supplier countries domestic governments are overwhelmingly the single greatest customer for national production and the single greatest sponsor of research and development. Supplier governments

... determine technical progress, and the size of ... defence industries, their structure and entry conditions as well as being able to regulate prices and profits. Of course, even for specialist military equipment produced domestically, the national government might not be the only buyer. It might allow foreign nations to purchase such equipment which is then reflected in the industry's export performance.[24]

From a supplier perspective therefore, exports do not represent a central position *per se*, but can be characterized as a secondary but important component of national security policy. If it is true that weapon system development is partly a function of an imperative to market equipment beyond the national armed forces, it is equally true that there are limits to the influence customers can be allowed if equipment is to be adequate in meeting the national requirements of major powers chiefly concerned with the threat posed by one another.

There is also a tension between the idea of a competitive market orientation in the naval arms trade, on the one hand, and the tendency towards long-term supplier–client relationships on the other. A relatively large group of naval contractors exist which are capable of meeting some of the needs of any naval programme, especially given that the size of many naval platforms increases the likelihood that they will contain a high percentage of imported sub-systems. At the same time, relationships between governments and particular suppliers have historically tended to be close and durable. The relationship often becomes so close and meshed that it is difficult for recipients to take their custom elsewhere or for suppliers to refuse equipment requests from important customers.

The contradictions within the naval arms trade reflect the growing complexity of the defence-industrial process itself. The nature and flow of military technology are changing in ways that may further undermine the traditional pattern of defence production and the arms trade. In the area of naval weapons there has been a limited movement towards collaborative production of weapon systems which were previously produced nationally. The most visible manifestation of this in the naval area is the programme to develop a NATO frigate and the weapon systems to arm it. While the exact shape of future collaborative weapon programmes and their political implications remain hazy, the tendency towards collaboration in itself represents a real change in the structure of naval defence production. Moreover, whatever the precise nature of shifts in military production, it is clear that within NATO, at least, there is a political head of steam behind the idea of both trans-Atlantic and intra-European collaboration in various forms of weapon procurement.[25] The increasing proliferation of binational and multinational systems produced in Europe will in turn change the nature of arms export policy. The export policies of suppliers have been under close national control, but the sale of multinational systems requires a different structure for managing transfers.

Collaboration and co-operation need not be confined to allies. The further growth of linkages between defence production in major powers and among developing countries is another probable development. Predictions that developing countries could develop fully autonomous arms industries have not been borne out. However, it is clear that a growing number of countries—in particular the newly industrializing countries (NICs) of South-East Asia—have the industrial capacity to contribute to the supply of components for developed countries, while European and North American companies in particular may desire cheaper labour costs to reduce unit costs and increase competitiveness. As a result, the relationship between military production in the industrialized world and the procurement of military equipment by developing countries is likely to become increasingly complex. This growing complexity will take place against a backdrop in which previously useful distinctions between First and Third Worlds are also losing their clarity. Mention has been made of the industrial capacity in some of the Asian NICs. Countries such as Brazil, China, Indonesia and India have, by dint of their regional military power and economic potential rather than their productivity, become increasingly significant international actors.

The pressures to collaborate in production

Many of the possible changes in the structure of the arms trade and arms production issue from the fact that the total volume of military production among the major powers will contract as these countries find it increasingly difficult to win domestic political support for large-scale military expenditure.[26] This process may be accelerated as a consequence of arms control agreements, although this seems less likely in the naval area than for ground forces. One consequence of this reduction in domestic demand has been a greater effort to export military equipment.[27] This has been a recent feature in several European countries, notably France and the United Kingdom, where there have been government-led military export drives. There have been one or two large orders from a few developing countries, but overall the effort has borne relatively little fruit in terms of new orders.[28] Moreover, several of the agreements which have been signed have subsequently been cancelled or significantly modified.[29]

For recipient governments (who set the parameters within which the arms trade operates since their policy defines demand for military equipment), the naval arms trade represents a complex series of choices concerning what to buy, how much and from whom. The process is dependent on a decision by the national political authority on naval tasks appropriate for the armed forces in the light of perceived threats and responsibilities. Consequently, programmes will be shaped by factors that include the nature of a country's foreign relations, doctrine, budget priorities, resource constraints, competing domestic objectives, the industrial capacity of possible suppliers, and the programmes and perceived future intentions of third parties. However, within this mixture of factors those of budget priorities and resource constraints have loomed large for many importers. In 1987, the United

Nations Report on Revitalizing Development, Growth and International Trade reported 'large increases during the early 1980s in the number of countries experiencing negative growth in expenditure . . . 18 of the 24 countries in Latin America and the Caribbean suffered declines in 1980–83. Among the 47 countries of Africa (other than North Africa) growth during this period was negative for 30.'[30]

Increased orders from some developing countries could have a significant impact on the defence industrial process in one or two Western countries and in the Soviet Union. However, on balance it seems unlikely that there will be a major increase in demand from developing countries. At least as significant are likely to be changes in the nature of relations between countries that are allied with one another but at different stages of economic and/or political development. The Southern Flank NATO countries (in particular Greece, Spain and Turkey), are very significant customers for naval equipment. These countries insist on a major domestic input to procurement programmes. Moreover, some of the costs of these programmes are met by multinational alliance funds.

Simply laying out these issues illustrates the immense complexity of managing a naval establishment and implies that changes in the level of naval forces cannot be brought about in the short term. This is not to say that the balance of naval forces is static, but simply that significant additional naval capabilities cannot be added overnight or even within a year.

The changing nature of weapon acquisition

For most countries the procurement of military equipment and the arms trade are intimately connected since, in the absence of national defence industries, the international market represents the greatest single source of arms and military equipment. However, there is evidence that the nature of international defence production is changing with important consequences for the arms trade. For its own purposes, the Independent European Program Group (IEPG) drawn from European members of NATO has sub-divided the procurement process into five basic categories, and in a modified form these also pertain for our purposes in considering naval procurement.[31] The IEPG classifies programmes into co-development, co-production, trade, licensed assembly and technology transfer. All of these processes involve a degree of technology transfer, which is less a means of procurement than a general rubric under which to discuss the subject. Co-development and co-production involve the largest degree of technology exchange. Co-development involves the joint design and development of a weapon system. Co-production means that two or more countries agree to produce a system designed and developed by a single country. Co-development would require a very intimate relationship between the agencies involved in development to the extent that companies involved may merge or become institutionally linked through a multinational organization. Developments of this kind have so far largely been confined to aerospace and missile programmes.[32] However, there are also naval programmes which have been identified as a potential focus for this type of collaboration. Co-production is in theory different from co-development in that it need not require the creation of multinational entities, but could work simply through the co-operation of national companies in multinational consortia.

Theoretical distinctions between these different processes have proved very difficult to sustain. Recipient companies are likely to modify or further develop imported designs to meet their own needs. Co-development involves the joint design and development of a weapon system, and as such does not fit any traditional definition of arms transfers. However, as an idea, co-development is of enormous potential significance for the arms trade. In future, systems developed and produced jointly could replace systems previously traded, in particular between members of the major alliances. If this were to happen it would have profound implications both for relationships between European countries and for the relations of those countries with the United States and Canada.

Genuine co-development would alleviate the recurrent problem of the imbalance in intra-alliance arms trade that sees the United States win the lion's share of competitive orders, and it is this prospect that has sustained the idea. The implications are particularly important for very large programmes, where the cost of research, design, development and production is simply beyond the budgetary possibilities of a single country other than one or other superpower. If the forum for organizing procurement of this kind was to be the IEPG within NATO, it would further contribute to the perception of a growing polarization between Europe and the United States as separate decision-making centres within the alliance. If the organizational setting for greater collaboration was to be outside the alliance, it would contribute to the perception in the United States and elsewhere that growing political unity in Western Europe was creating a powerful economic bloc.

Licensed assembly is a limited form of co-operative production in which one country sells another a package of data and skills, along with the right to assemble a given weapon system supplied to them in kit form. Trade involves the import or export of finished weapons. These methods of procurement are discussed in part II, illustrated by reference to naval programmes currently under way in selected countries, to determine the degree to which: *(a)* the acquisition of naval equipment imposes a logic on the procurement process; and *(b)* the procurement process responds to specific economic, geographic and historical conditions.

Notes and references

1   In *The Jellicoe Papers: The Private and Official Correspondence*, vol. 1 (The Navy Records Society: London, 1966), pp. 26–27.
2   This lacuna has been addressed in recent years by Morris, M., *Expansion of Third World Navies* (Macmillan: London, 1987). Other authors have discussed international transfers as they affect the shipbuilding industry: Todd, D., *The World Shipbuilding Industry* (Croom Helm: London, 1985); and Faltas, S., *Arms Markets and Armaments Policy: The Changing Structure of Naval Industries in Western Europe* (Martinus Nijhoff Publishers: Doordrecht, 1986).
3   'Shipbuilding deal', *Jane's Defence Weekly*, 14 May 1988, p. 948; Ciampi, A., 'Italian Naval Defence Industry', *Navy International*, Sep. 1988, pp. 440–46; *Defence & Armament Heracles*, Dec. 1988, p. 30.
4   Ciampi (note 3).
5   Poland builds two classes of amphibious ships for the Soviet Union, the Ropucha class and the smaller Polnochny class. Both were designed by the Soviet Union. See Carolla, M. A., Herries, J. P. and Good, C. W., 'Amphibious ships, mine warfare ships, corvettes, and missile patrol and torpedo craft', in B. W. Watson and S. M. Watson (eds), *The Soviet Navy: Strengths and Liabilities* (Westview: Boulder, Colo., 1986). In addition, Poland is building a new class of Signals Intelligence ships for the Soviet Navy called the Vishnya class, *Soldat und Technik*, Dec. 1987, pp. 730–34.
6   Heisbourg, F., 'Public policy and the European arms market', in P. Creasey and S. May (eds), *The European Armaments Market and Procurement Cooperation* (Macmillan: London, 1988), p. 86.
7   A conceptual framework for looking at these processes is contained in MccGwire, M., 'The horizontal proliferation of maritime weapons systems', in G. H. Quester (ed.), *Navies and Arms Control* (Praeger: New York, 1980).
8   The text of the Law of the Sea (LOS) Convention is reproduced in *International Legal Materials*, no. 21, Nov. 1982, pp. 1261–354.

9   Part V of the Law of the Sea Convention is reproduced in Westing A. H., *Global Resources and International Conflict*, SIPRI (Oxford University Press: Oxford, 1986), pp. 233–60.

10   *Asian Recorder*, 16–22 July 1989, p. 20 682.

11   For a discussion of the impact of naval arms control measures, see Fieldhouse, R. and Taoka, S., SIPRI, *Superpowers at Sea: An Assessment of the Naval Arms Race*, Strategic Issue Papers (Oxford University Press: Oxford, 1989).

12   Ranft, B., 'Restraints on war at sea before 1945', in M. Howard (ed.), *Restraints on War: Studies in the Limitation of Armed Conflict* (Oxford University Press: Oxford, 1979), p. 42.

13   Modelski, G. and Thomson, W. R., *Seapower in Global Politics 1494-1993* (Macmillan: London, 1988).

14   Faltas (note 2), pp. 97–99.

15   See especially Faltas (note 2).

16   Especially Morris (note 2); and in many articles. See also Janis, M. W., *Sea Power and the Law of the Sea* (Lexington: Mass., 1976); and Larson, D. L., 'Naval weaponry and the law of the sea', *Ocean Development and International Law*. vol 18, no. 2, 1987.

17   Hill, J. R., *Maritime Strategy for Medium Powers* (US Naval Institute Press: Annapolis, 1986), p. 14.

18   Till, G., *Maritime Strategy and the Nuclear Age* (Macmillan: London, 1984), p. 13.

19   Especially Morris (note 2); Karkoska, A., 'The expansion of naval forces', *World Armaments and Disarmament: SIPRI Yearbook 1979* (Taylor & Francis: London, 1979), pp. 357–74; Karkoska, A., 'Naval forces', *Ocean Yearbook 2* (University of Chicago Press: Chicago, 1980), pp. 199–226; Janis, M. W., *Sea Power and the Law of the Sea* (Lexington: Mass., 1976); and Larson, D. L., 'Naval weaponry and the Law of the Sea' *Ocean Development and International Law*, vol. 18, no. 2, 1987.

20   The best known of these works are probably Cable, J., *Diplomacy at Sea* (Macmillan: London, 1985); Cable, J., *Gunboat Diplomacy* (three editions), Kelleher, C. and Booth, K., *Navies and Foreign Policy* (Croom Helm: New York, 1979); and Booth, K., *Law, Force and Diplomacy at Sea* (Allen & Unwin: London, 1985).

21   For a discussion of these efforts, see Hu, N. T. A. and Oliver, J. K., 'A framework for small navy theory: The UN Law of the Sea Convention', *Naval War College Review*, spring 1988, p. 39.

22   United Nations study series on disarmament, no. 16, *The Naval Arms Race* (United Nations: New York, 1986), p. 71.

23   Faltas (note 2), pp. 57–58.

24   Hartley, K., 'The European Defence Market and Industry', in P: Creasey and S. May (eds), *The European Armaments Market and Procurement Cooperation* (Macmillan: London, 1988), p. 32.

25   The RAND Corporation is currently engaged in a multi-year study of the potential for collaboration of various kinds in this sphere. The 'Two Way Defense Trade' project is sponsored by the Office of the Under Secretary of Defense for Acquisition.

26   Discussed in Deger, S., 'World military expenditure', *SIPRI Yearbook 1989: World Armaments and Disarmament* (Oxford University Press: London, 1989), pp. 131–171.

27 Brzoska, M., 'The future of arms exports from Western Europe: a case of forced conversion?', *Institut für Politische Wissenschaft*, working paper no. 26, 1988.

28 The case that arms transfer policy has become increasingly commercialized is argued by Karp, A., in 'Ballistic missile proliferation in the Third World', *SIPRI Yearbook 1988: World Armaments and Disarmament* (Oxford University Press: London, 1988), pp. 287–310.

29 These include British deals with Jordan, Malaysia and Oman. For an overview of developments in 1988, see Anthony I., in *SIPRI Yearbook 1989* (note 26), pp. 195–269.

30 Quoted in Brzoska (note 27).

31 For a discussion of the IEPG and its function see Covington, T. G., Brendley, K. W. and Chenoweth, M. E., *A Review of European Arms Collaboration and Prospects for its Expansion under the Independent European Program Group*, RAND-N—2638-ACQ, July 1987. The RAND note contains no discussion of naval programmes, but is a succinct overview of the history of the US and NATO European efforts in collaborative arms procurement.

32 See for example, Edmonds, M. (ed.), *International Arms Procurement: New Directions* (Pergamon: New York, 1981).

2. The naval market

I. Introduction

The aggregate size of the trade in naval weapon systems and platforms between 1979 and 1988 is valued at US $75 819 million.[1] In other words, during this period the naval arms trade has accounted for roughly 22 per cent of the total arms trade by value (see table 2.1). Moreover, this figure excludes land-based fixed-wing aircraft which have anti-ship or anti-submarine capabilities but are not exclusively dedicated to these missions.

Table 2.1. The naval arms trade within the total arms trade 1979-88
SIPRI trend indicator values in US $ million (1985)

	Total arms trade	Naval sales	Naval equipment as % of total
1979	32 054	5 488	17
1980	30 079	7 749	26
1981	35 212	9 101	26
1982	34 218	8 588	25
1983	33 208	8 419	25
1984	34 112	8 927	26
1985	32 284	5 875	18
1986	34 647	7 007	20
1987	39 518	7 243	18
1988	33 969	7 422	22
1979–88	339 301	75 819	22.3

Within this aggregate figure, it is possible to identify with some precision the important actors, both as sellers and buyers of naval equipment.[2]

The largest suppliers of warships and auxiliary vessels are the Federal Republic of Germany, France, Italy, the Soviet Union, the United Kingdom and the United States. These six countries among them account for roughly 86 per cent by value of the hulls sold during the decade 1979–88.

From the point of view of recipients, there is a clear regional dynamic within the naval arms trade. Regional focal points are South-East Asia, the South Atlantic and the Mediterranean. The ASEAN region in particular contains some of the principal naval importers, while North and South Korea, India and Australia lie in contiguous regions. The list of major customers contains eight 'pairs' of close or direct neighbours—Argentina/Brazil; North Korea/South Korea; Indonesia/Australia; India/Indonesia (through the proximity of India's Andaman and Nicobar island possessions); Indonesia/Thailand; Singapore/Indonesia; Syria/Turkey; and Greece/Turkey.

While the pattern is fairly clear, it would not be appropriate to overstress the regional dynamic. It is partly a simple function of the fact that navies operating in close proximity to one another operate broadly similar vessels because of the needs imposed by their maritime geography. The surface component of Latin American navies of the southern cone and those of the North Atlantic region, for example, are composed of vessels of higher tonnage than those of Middle East navies. Sea conditions are very different in the South Atlantic as opposed to the Mediterranean or Persian Gulf. Geography has a powerful impact within regions as well. In Europe, the Baltic littoral states operate with shorter vessels and require only limited endurance compared with the British or Dutch navies, which have to operate in the North Sea and/or the North Atlantic. This has contributed to a requirement for ships of smaller tonnage.

Another clear feature of the structure of the market is the link between bilateral security relationships and the transfer of naval technology. This is a feature that transfers of naval technology share with the wider arms market, which is dominated by a relatively small number of arms relationships between countries which are either formal allies or linked in some form of bilateral treaty system. Looking at the exports of British, Soviet and US warships in appendix 1, for example, reveals that the most important customers have been as follows. For the UK, recipients of capital ships, in particular, have been former colonies which either are or have been members of the Commonwealth, such as Australia, India, Malaysia, New Zealand and Pakistan. In the case of the USA, the most important customers have been European allies, Latin American countries, especially Argentina and Brazil, and Asian countries to which the US has or has had formal security commitments, such as Pakistan, South Korea and Taiwan. For the Soviet Union the most important customers have been allies—such as Bulgaria, Cuba, the GDR, Poland and Romania—and friendly developing countries such as Egypt before 1974, Indonesia before 1965, North Korea, Syria and Viet Nam. This feature of transfers is also apparent in the case of smaller exporters. For the Netherlands the most important customer has been a former colony, Indonesia. China has exported most to Bangladesh, North Korea and Pakistan, with which countries it has a history of close relations.

II. The sellers

Sales of new vessels

Table 2.2 offers some greater clarity about the most important exporters of warships. While the Soviet Union and the United States are the two largest exporters, the table indicates the relative importance of European suppliers. The Federal Republic of Germany, in particular, has a prominence in this area that it does not display in other areas of the arms market, ranking third in the aggregate value of exports 1979-88. The nature of exports by these suppliers is discussed in greater detail below.

Table 2.2. Major suppliers of ships as a percentage of total ship transfers, 1979–88

Supplier	USSR	USA	FRG	France	UK	Italy	Total
Percentage	20	18	16	12.5	12	8	86.5

Table 2.3 below lists the major suppliers of naval equipment according to the definition described in note 1. The inclusion of naval aircraft and weapon systems dilutes the importance of the Federal Republic of Germany and increases the importance of France, the United States and, to a lesser extent, Italy as suppliers. For the United States this mostly reflects the widespread export of naval guided missiles, especially various naval versions of the Standard and Sparrow anti-aircraft missiles and the Harpoon anti-ship missile. It also reflects the enormous unit cost of certain specific aircraft, notably the P-3C Orion in its maritime patrol versions. In the case of France, the widespread export of naval helicopters and the Exocet family of

missiles adds significantly to the value of naval exports. The values for French and British sales represented here are complicated by the fact that both countries have produced naval aircraft collaboratively. Sales of the SA-341 Gazelle helicopter are attributed to France in spite of considerable involvement by the British company Westland because they are programmes under French leadership. Sales of the Lynx helicopter are attributed to the UK although Aérospatiale of France builds 30 per cent of the airframe.

Table 2.3. Value of exports of naval equipment by major suppliers[a]
Values are in US $ million (1985)

	USA	France	USSR	FRG	UK	Italy
1979	1 055	1 352	1 317	250	413	909
1980	2 847	1 799	1 324	506	523	529
1981	2 585	2 166	716	1 010	866	1 074
1982	2 358	2 245	1 255	236	1 256	839
1983	3 024	2 197	1 323	1 137	202	682
1984	1 962	2 409	875	1 905	805	552
1985	2 005	2 631	832	297	351	289
1986	2 229	2 648	1 359	516	394	253
1987	2 787	1 598	1 246	458	972	141
1988	4 004	1 413	960	1 018	399	327
1979–88	24 856[b]	20 991	11 207	7 766	6 376	5 536

[a] According to the definition in note 1.
[b] Figures may not add up due to rounding.

The sellers of naval systems are of different types and can be sorted into three categories: sellers of major, new naval platforms; sellers of major, second-hand naval platforms, and sellers of naval equipment. These categories are not mutually exclusive, and it would be possible to belong to all three.

A relatively small group of countries is able to sell major new surface ships. The group of suppliers is confined to European naval shipbuilders, the United States and the Soviet Union, of which European countries are collectively the most important suppliers. The sale of complete, new naval vessels was historically a North–South trade, and European shipbuilders have products that are better suited to customers with limited resources. France, FR Germany, Italy and the Netherlands, in particular, have developed surface ships of a size and design attractive to developing countries.[3] FR Germany and the

Soviet Union have come to dominate the export of submarines. In the 1980s, Greece, Spain and Turkey, who built their navies on second-hand ships after 1947, have become important customers for new ships either produced by one of the major naval powers or designed there and constructed under licence by the recipient.

There are countries which have the industrial capacity and design skills to produce and sell new ships but do not, notably China, India and Japan. None of these is a major exporter of new vessels, but for very different reasons. Japan has a shipbuilding industry capable of producing a wide range of modern ships suited for many different missions. However, Japan is restricted in what it can sell by the interpretation successive governments have placed on provisions in the Constitution. Japan has on occasion built vessels for foreign coast guards, notably the Philippines, and transferred technology to Third World shipbuilders including India and Singapore. China has exported new vessels of smaller tonnage and versions of the Soviet Romeo class submarine, but for the most part new ships have been built for the People's Liberation Army (PLA). Similarly, Indian shipbuilding is oriented towards meeting the demands of the Indian Navy and Coast Guard.

Some other countries may emerge as naval suppliers on a small scale. Romania has recently begun to design and build larger naval vessels, but as yet has no record as a seller of these vessels. A South Korean shipyard was one of four suppliers that submitted designs to Taiwan for a new frigate, although the Ulsan design was not short-listed. As a consequence of resource constraints rather than a change in policy, Argentina may be forced to sell two of four MEKO-140 frigates produced under licence from FR Germany.[4] Peru, a significant importer of Soviet equipment, has been reported as having contracted to construct as many as 80 ships for the Soviet Union as part of an effort to refinance Peru's debt to Moscow. While most of these vessels would be civilian, the plan apparently includes 20 light troop carriers and 10 Coast Guard vessels.[5] New suppliers are more active as sellers of smaller surface vessels. However, even in this sphere, potential exporters among the developing countries are heavily dependent on foreign warship design and technology. This is true of Singapore, where two shipbuilding companies in particular— Vosper Singapore and Singapore Shipbuilding and Engineering— have exported successfully to South-East Asia, Taiwan and the smaller Persian Gulf states. These companies have bought British and

West German designs for fast attack craft (FAC), landing ships and patrol craft.[6] This is also the case for some European producers, notably Spain, which have imported proven design know-how to support established shipbuilding industries without incurring the costs of research, development and trials. All of the major naval vessels built in Spain were designed elsewhere. The aircraft-carrier *Principe de Asturias* and Santa Maria class frigates are of US design, while other frigates are of Portuguese and West German design.[7]

Of the countries that have sold new vessels, some are restricted in their export policy by the nature of their national demands. It is clear from table 2.1 that the United States is not an insignificant supplier of warships, aggregate sales from 1979–87 placing it in second place in the table of suppliers. In particular the FFG-7 Oliver Hazard Perry class has become a successful export design. However, as noted above, the US is not the dominant force in this area of the arms trade that it has been elsewhere. In the case of the United States, very few of the major naval vessels in production would be available for export without modification. Many contain sensitive technology and/or nuclear propulsion systems that cannot be transferred as a result of different US export regulations. Moreover, of those available for export, few would be affordable for potential buyers. This group of potential buyers would in itself be a small number of countries given the technological problems of operating systems of this kind. The unit cost of an air defence escort of the type being produced by the United States would approach $1 billion. This applies to a lesser extent to Canada, where ships built have also tended towards high tonnage and sophisticated (imported) equipment fits, and to the United Kingdom. The designs currently under construction in the UK, Type 42 destroyers, Type 22 and Type 23 frigates, are all of 3500 tonnes displacement or more, while the programme unit cost of a Type 23 frigate is in the region of $270 million (in 1987 prices).[8]

The Soviet Union is the largest single exporter of ships. Arms export patterns have changed along with other areas of Soviet naval policy. In the 1960s, the Soviet Union was a very successful exporter of smaller patrol craft and FACs. Examples of exports of submarines and major surface ships were rare, although they did occur, notably the sale of a Sverdlov class cruiser to Indonesia and the sale of Skory class destroyers to Egypt and Indonesia in the 1960s.[9] During the 1950s and 1960s, the emphasis of Soviet policy shifted to the production of large numbers of submarines, and during the 1970s these vessels were sold to Soviet clients. Soviet shipbuilding policy underwent a further change in the 1970s, with the tendency to build much larger surface vessels that tend to carry formidable quantities of armament. Vessels of this kind have been sold in quantity only to one close Soviet arms client with a large naval establishment, India. These ships would be impossible for small navies to man and maintain, but there are other factors which have restricted Soviet exports. One limiting factor has been the relatively small naval shipbuilding capacity of the Soviet Union not in absolute terms but relative to Soviet programmes. The decision to construct a very large submarine fleet and expand the surface component of the Soviet Navy in the 1960s stretched shipbuilding capacity. There is also evidence of Soviet efforts to match exports to their own immediate interests as well as meeting the requests of important clients. In the cases of Egypt and Indonesia, the transfer of submarines and FACs was more likely to complicate US naval operations in the context of the regional maritime geography than the sale of large warships, although the latter were persistently requested, at least by Egypt.[10] In the Indian Ocean context however, the opposite judgement applied.

The Federal Republic of Germany is a supplier of new warships on the same scale as the superpowers. The FRG sells ships to a range of countries—ships have been delivered to 15 different navies since 1979—but the bulk of sales is directed to a much smaller group of clients, among which Argentina, Greece and Turkey are the most important. These are customers with which the FRG has developed an intimate and long-standing supply relationship, and West German vessels tend to be successful in winning follow-on orders from clients. West German success as an exporter has been a product of several factors which have, in concert, shaped a naval industry ideally suited to the current needs of some important customers. Paradoxically, this is in part a product of Western European Union (WEU) regulations which originally disallowed the building of ships in excess of 3000 tonnes for the West German Navy.[11] These regulations were designed to handicap the development of a major naval shipbuilding capacity in the Federal Republic in deference to the concerns in other European countries about German rearmament. As a result, West German shipyards were forced to specialize in designing and building ships capable of meeting the exacting demands of the European military environment but of a size which reduced both unit production cost and subsequent running costs—in particular manning levels. The lack of a domestic production base for weapon systems and electronics was also in part a function of the post-war political environment. However, the lack of home-made weapon systems forced the naval industry to become expert in integrating a wide range of systems of different national origin into West German hulls. The organizational skills required to manage an integrated programme of this kind have had a direct relevance in the area of naval exports, where procurement programmes often involve several companies' from different countries supplying equipment before a naval platform is commissioned.

Domestic politics have also contributed indirectly to West German export success, in that orders for ships for the West German Navy have tended to be spread between several yards in part to preserve employment and in part to force down production costs through competitive tendering.[12] A result has been that customers have found that sufficient capacity exists to allow more than one ship to be laid down simultaneously, shortening the overall time for the programme. Moreover, the Government has framed its arms export policy to ensure that a broadly based domestic naval industry is sustained. When, in 1977, the Blohm and Voss shipyard lost a tender to supply frigates to the Navy, export orders for the unsuccessful MEKO-360 class design to Argentina were not opposed by the Foreign Office.[13] The interpretation of arms export guidelines by Foreign Office officials have been helpful in that Argentina, Iran and South Korea have been identified as suitable customers for submarines at various times since 1979.

Among exports of warships from FR Germany, submarines have come to play an increasingly important role. In the 1980s the navies of Argentina, Brazil, Chile, Greece, India and Peru have all received submarines designed by the Federal Republic. Norway and Israel are awaiting submarines from the FRG, and Denmark will buy submarines of West German design from Norway once they have been refitted.[14] The ability of FR Germany to build submarines has been sustained entirely by export orders in the 1980s. The last submarine built for the West German Navy was completed in 1979, and there are no plans for new construction before 1991 at the earliest. Moreover, export orders have allowed the development of design skills, notably by the Ingenieurkontor Lübeck company. Whereas the submarines in service with the West German Navy are of less than 500 tonnes displacement and use diesel engines, the submarines exported to Argentina, Brazil and India are approaching 2000 tonnes and use more sophisticated propulsion technologies. These enhanced design skills are to be reflected in the new generation of submarines for the West German Navy.[15] The growing export of submarines from the Federal Republic has also reflected the fact that other countries within the Western alliance produce submarine designs that would be difficult for less developed navies to man and maintain. France and the United States have no conventional submarines currently under construction, while the UK produces only the Type 2400 Upholder class, which displaces over 2400 tonnes and costs in the region of $330 million to build (1987 prices). As a result, countries which require a conventional submarine of recent design and cannot or will not turn to the Soviet Union or China have few options but to buy from FR Germany, Italy, the Netherlands or Sweden.

Second-hand vessels

According to one estimate, second-hand warships still represent the majority of major warships traded between countries.[16] Clearly this remains an important part of the market.

Appendix 1 lists all second-hand ships transferred since 1947. Second-hand sales and leases have in the past been the means by which major naval powers disposed of surplus vessels. The United States, the Soviet Union and the United Kingdom have dominated the transfer of such vessels during the post-World War II period, with the UK being significantly less important than the superpowers. The United States has transferred well over 1000 hulls and the Soviet Union around 1000 during this period. As noted above, the recipients of these vessels have overwhelmingly been the navies of allies or friendly powers. Moreover, many of these transfers have been either free—in the context of military aid programmes—or on nominal financial terms.[17] European countries which are now first-ranking naval shipbuilders bought significant numbers of second-hand US ships during the 1950s and 1960s. This was especially true for France, FR Germany and Italy, although in the case of FR Germany this process also involved the return of German ships impounded after World War II. During the same period Bulgaria, the German Democratic Republic, Poland and Romania were in receipt of large numbers of vessels from the Soviet Union.

While for the largest naval powers disposing of older ships has taken place as part of a process of modernization, most Third World navies have been built around second-hand ships. Navies in which second-hand ships have historically been virtually the only kind in service include Colombia, Egypt, Pakistan, Peru, South Korea, Taiwan and Uruguay. In many cases these transfers involved ships built during World War II which remained in service until the 1970s.

There has been a shift in the nature of the market for warships in the period since 1980 away from the procurement of second-hand ships towards the purchase or construction of new hulls. Certainly, some countries which traditionally relied on buying second-hand ships from major powers, notably Greece, Spain and Turkey, now build new ships. However, as a general statement it may be premature to suggest that second-hand ships are losing their importance. In the 1980s, Greece and Turkey have been major recipients of second-hand US vessels (see appendix 1) as well as ordering new construction from FR Germany. Among Third World navies there has been a heightened interest in US Navy and Royal Navy vessels that have become available for export. During the late 1970s the United States was forced to confront a problem of block obsolescence within its surface fleet which, combined with the naval policies pursued under the Reagan Administration, led to an acceleration in the laying down of new hulls.[18] As these vessels enter service, there may be an increasing number of second-hand US escort vessels added to the disposal list (as opposed to the naval reserve). There is a clear market for these ships among countries with a historically close security relationship to the United States, notably Brazil, Pakistan, the Philippines and Taiwan. All of these countries, and South Korea, registered an interest in 16 frigates which were decommissioned from the US Navy in 1988.[19] Whether these ships eventually become available for sale depends on US budgetary considerations. The US Navy would have preferred the 16 frigates mentioned above to enter the naval reserve. However, the budget agreement between Congress and the Administration eventually resulted in cuts of $12.3 million to be absorbed by the US Navy and Marine Corps, as a partial consequence of which these ships entered the disposal list.[20] As a compromise, these ships have not been sold but have in the short term been leased to countries with close arms relationships with the United States. Pakistan and the United States agreed in August 1988 on the lease of eight of these vessels, four Brooke class and four Garcia class frigates, together with a 1940s vintage repair ship. All of the vessels were transferred with all of their anti-submarine warfare systems but without any air defence systems.[21]

Sellers of naval sub-systems

The relationship of sellers of sub-systems to the naval market is the most complex. In spite of their name, many of these sub-systems are neither simple nor insignificant in themselves. At the furthest end of the spectrum, the term can refer to an integrated suite of sensors, weapons and communications systems of enormous complexity and with a very high unit cost. On 24 June 1988, Japan and the USA signed a contract for the sale of the Aegis, a computer-controlled air defence system capable of intercepting 10 or more targets (including anti-ship missiles) simultaneously.[22] The Aegis air defence system incorporated on US vessels has been valued at $486 million, and while the version sold to Japan is less elaborate, it is clearly enormously expensive. The difference in the versions lies in the radar and data processing elements rather than the weapon systems themselves. The Japanese version incorporates a SPY-1D radar with a network of UYK-43 computer systems while the United States version includes the SPY-1A with the UYK-7 computer system.[23]

The Japanese decision to purchase Aegis highlights the fact that sales of sub-systems as much as other arms deals can become highly politicized. The prospective sale of the Aegis, to be installed in a new 7000-tonne air defence escort, had been discussed by the two governments since 1984, but US congressional opposition culminated in the approval on 3 March 1988 of a motion by the Seapower Subcommittee of the House of Representatives to prohibit the sale.[24] The concerns expressed by the Subcommittee related to safeguarding US technology from potential enemies in the wake of the illegal export of four computerized milling machines to the Soviet Union by the Toshiba Corporation of Japan.[25] The Subcommittee's decision did not prevent the offer of a contract to Japan with regard to the sale, but in June a group of Congressmen attached an amendment to the Fiscal Year 1988–89 Defence Appropriations Bill that would require all buyers of the Aegis to buy US-built ships in which to install them.[26] In response, Secretary of State George Shultz and Defense Secretary Frank Carlucci co-authored a newspaper article in which they advocated the sale, and in August 1988 it was confirmed.[27]

Apart from the size and sophistication of some naval sub-systems, their transfer also illustrates the great complexity of some naval arms deals. Mention has been made of the fact that major surface ships contain many different systems integrated with one another on the same platform. The procurement of ships therefore can require the creation of multinational corporations to manage programmes and this in itself carries important consequences, such as the advantages that FR Germany has drawn from its particular organizational and management skills. To illustrate this point, when Turkey took possession of the first of four MEKO-200 frigates on 17 July 1987, the Federal Republic of Germany (specifically, Blohm and Voss shipyard) was ostensibly the supplier. However, the combat systems on board incorporated equipment from the FRG, the Netherlands, the United Kingdom, the United States (mainly) and Switzerland.[28] MEKO-200 frigates sold to Greece will include US sonars, communications systems, torpedoes and anti-ship missiles worth roughly $150 million.[29]

Dependence on foreign companies for the management of complicated programmes is not restricted to less developed countries. In Canada the management of the programme to produce City class frigates is conducted by a US firm, Sperry. In Australia neither Australian industry nor the prime Swedish contractor Kockums felt able to undertake overall programme management without assistance from Bath Iron Works of the United States as consultants.

The Turkish MEKO-200 programme also illustrates the impact of the time that complicated naval programmes take to fulfil. The contract between Turkey and the Federal Republic was signed on 29 December 1982 for four frigates, the last of which is expected to be delivered in 1989.[30] Moreover, this schedule of delivery—four vessels within seven years—is by no means unusual. In fact, this delivery schedule was possible only because production was split between three shipyards, two in the Federal Republic and the Gölcük yard in Turkey.

Shipborne aircraft are also sub-systems of a sort, and a very limited group of manufacturers produces specialized equipment. Only the Soviet armed forces have bought Soviet fixed-wing aircraft for deployment on an aircraft-carrier. Currently, only France, the United States and the United Kingdom export this type of system. European navies, other than the Royal Navy, have relied on imported aircraft for their aircraft-carriers, including the French, whose aircraft-carriers *Foch* and *Clemenceau* each carry 10 F-8E Crusaders bought from the United States in the mid-1960s.[31] Spain faced a choice between two different versions of the Harrier aircraft available from the UK and the US, opting for the US version, the Harrier AV-8B. In February 1989 the Italian Parliament approved a change to the law allowing the Navy to operate fixed-wing aircraft, and the British version of the Sea Harrier is believed to be the favoured choice to equip the helicopter carrier *Giuseppe Garibaldi*. Sea Harriers conducted trials from the *Giuseppe Garibaldi* in May 1988. In February 1989 Aeritalia signed a letter of intent and a Memorandum of Understanding with British Aerospace for joint development of a new version of the Sea Harrier which was expected to lead to the purchase of 12–18 aircraft and possibly further collaborative projects between the two companies.[32] The Italian and Spanish decisions to develop their maritime air forces with different versions of the same aircraft are of great potential significance in the context of growing discussions among Mediterranean littoral countries concerning their regional security. In July 1988, Italy and Spain signed a bilateral agreement in Rome stating that they would co-ordinate their air defences and provide for greater technical understanding between their navies. Italian Minister of Defence Valerio Zanone has identified the maritime defence of Italy as one of the central pillars of Italian defence policy to 2000.[33]

A larger group of countries is able to supply helicopters, although within each of these supplier countries the number of companies making helicopters has declined. In 1986, there were 18 companies able to build helicopters in the United States, 2 in France and 5 in the United Kingdom. By 1988, the number of companies building helicopters was 10, 2 and 2, respectively.[34] France, the United States and the United Kingdom, in particular, have made widespread sales of these systems. For France, exports of helicopters (both civilian and military) account for over 80 per cent of Aérospatiale's total turnover.[35] The UK and the USA have marketed different versions of the same aircraft, the Sea King, produced by Westland of the UK and under licence by Sikorsky of the United States.[36] The SH-60 Seahawk, also made by Sikorsky, is an increasingly successful helicopter measured in terms of export sales.

III. The buyers

The clearest manifestation of the horizontal proliferation of naval systems is the growth in the number of naval forces in the world. For example, issues of *Jane's Fighting Ships* for the respective years list 67 navies in the world in 1958, 91 in 1968, 152 in 1978 and 152 in 1988. As the stabilization in the number of navies indicates, the number is directly linked to the process of decolonization and state formation within the international system. However, while almost all coastal countries may be participants in the naval arms trade the nature and scale of their participation vary enormously. The vast majority of these 152 navies are purely coastal navies, with little or no ability to operate beyond a narrow strip of sea space within 12 nautical miles of their coast.

By any criteria, other than domestic perception in smaller countries, the naval arms trade is dominated by a small number of importers. Consequently it is possible to identify with some precision the most important actors in the naval arms trade. Among Third World countries Argentina, Brazil, India, South Korea and Turkey have all committed themselves either formally or implicitly to large naval programmes since 1981. Chile, Indonesia, Iraq, Pakistan and Saudi Arabia will invest more modestly but still significantly in naval capabilities. While the naval policy of the United States has attracted the greatest attention, other industrialized countries have announced significant naval programmes, notably Australia (in conjunction with New Zealand in some programmes), Canada and Japan.

The cases examined more specifically in part II of this book are Argentina, Australia, Brazil, Canada, India and New Zealand. These countries are all among the important importers of naval equipment. However, other than a common emphasis on their maritime security, there is apparently little else which unites these countries. Canada and Australia combine low populations with a benign threat environment. In these characteristics they could hardly differ more dramatically from India. Australia and Canada also have advanced and relatively successful economies as compared to Argentina and Brazil, which face major economic problems. Australia and Canada are members of alliances which contain the United States. India on the other hand has been a central figure in the non-aligned movement and a frequent critic of US naval policy, particularly in the Indian Ocean.[37]

All of these countries are linked by the fact that they have recently undertaken major naval policy reviews which led to the announcement of programmes that would make a significant impact on the naval balance within their region. India has become the first ever customer for a nuclear-powered attack submarine (SSN) and Canada stated its intention to purchase up to 12 much more modern and capable SSNs. Argentina and Brazil have discussed the same kind of platform for their own navies, possibly incorporating indigenously designed nuclear propulsion systems. India has recently taken delivery of an ex-Royal Navy light aircraft-carrier, commissioned as the *INS Viraat*, not as a replacement to the aircraft-carrier in service but in addition to the *INS Vikrant*, which will now undergo an extensive life extension refit before returning to service. In 1983 the Australian Government decided not to replace the aircraft-carrier *HMAS Melbourne*, abolishing sea-based fixed-wing aircraft from the Australian Navy. In Argentina and Brazil decisions on the replacement of aircraft-carriers built in the 1940s have been deferred. In Australia and New Zealand naval programmes have become meshed with changes in attitude to alliance membership and new perceptions of regional security policy.

Importers of new ships

The trade in major new naval vessels is dominated by a relatively small number of recipients, few of whom could be called 'new' naval powers. These primary customers have naval traditions stretching back to the 1960s, often much further. Argentina, Brazil and Chile built their navies between 1910 and 1940 around new construction from Italy, Spain, the UK (in particular) and the United States.[38] In Argentina, Brazil, Chile, Greece, India and Turkey experience of large-scale naval operations goes back to the 1920s and 1930s, and naval personnel served in World War II. In India the naval tradition is older than the country itself. These traditional naval powers tend to include a balance of different systems in their navies. For example, Argentina, Brazil and India operate aircraft-carriers, and if the naval arms trade is confined to a relatively small number of important recipients, this is particularly true of the acquisition of new, large naval vessels. In the case of major new surface vessels, for example, Argentina, Brazil and India between them accounted for almost 30 per cent of the trade in vessels of 1500 tonnes or over between 1983 and 1987.[39] The other primary customers for these types of vessels are countries within the major alliances which do not themselves have a naval shipbuilding industry capable of meeting national requirements, in particular Australia, Greece, Spain and Turkey. In Spain, as noted above, the tendency has been to import proven ship designs and the expertise to build them in Spanish yards, notably the Bazan Shipyard. Recent examples of designs bought from abroad include the aircraft-carrier *Principe de Asturias,* which is based on a US design for what was to be called a 'sea control' ship rather than an aircraft-carrier. The design was rejected by the US Navy as being too small and limited to meet their requirements.[40]

In all of these countries it is clear that the interests of the domestic shipbuilding industry have a direct impact on procurement policy. All of these countries insist on a significant domestic contribution in the production process, and a pattern has emerged in which the direct import of one or two completed ships built in the shipyards of the exporter is followed by the production under licence of subsequent vessels in the shipyards of the recipient. The recipient country might have expertise in some areas which would enable it to contribute national elements to the completed vessels and is likely to want to use the programme to enhance these elements and also contribute to local employment.

The fact that important actors in the naval arms trade are spread among the developed and developing, allied and non-aligned world tends to reinforce the suggestion that structuring naval forces and acquiring the equipment for them demand a planning process which has a certain logic. However, the trade in second-hand equipment, the importance of which has already been noted, is by nature opportunistic rather than planned. Procurement programmes noted thus far have characteristically been planned meticulously and many years in advance. It is not possible to plan a programme in this way on the basis that systems *may* be retired by the navies of other countries at some distant future date, especially in the knowledge that such systems, where they are available, may be sought by several navies. From the discussion below, it emerges that the countries which tend to rely on second-hand ships are those where long-range economic planning is difficult.

Importers of second-hand ships

At times in the past all developing countries have relied on buying second-hand ships from major power navies. It is clear that this is still the preferred choice for some countries unable to release foreign exchange for naval programmes. This is particularly true of countries with a naval tradition that includes operating major surface vessels which now face the problem of block obsolescence in their fleets. The navies which face the most severe problems of obsolescence are Bangladesh, Brazil, Chile, Mexico, Pakistan, Peru, the Philippines,

South Korea, Taiwan and Uruguay. From appendix 1 it can be seen that, in many cases, these are the countries which have historically had the keenest interest in second-hand vessels that have become available. As noted above, these countries are in many cases either allies of the United States or countries with close bilateral relations. The navies of Pakistan, the Philippines, South Korea and Taiwan, in particular, fall into this category and have largely been constructed from ex-US Navy ships of various classes. It is also clear from appendix 1 that the naval programmes of countries forced to rely on second-hand vessels are dictated by the naval policies of the major powers. This is especially true for the United States. Of the escort vessels sold second hand, the Buckley class, Fletcher class, Gearing class and Allen M. Sumner class have been particularly widely sold. Buckley and Fletcher class convoy escorts were built quite literally in their hundreds between 1941 and 1945. In addition to over 200 in the US Navy, over 50 were transferred to the UK under Lend-Lease arrangements. These ships began to become available on retirement from the US Navy and the Royal Navy in the late 1950s and the early 1960s.[41] The Allen M. Sumner class is an enlarged version of the Gearing class and both were constructed in large numbers after 1946. These vessels became available on their retirement in the late 1960s and through the 1970s.

Buying second-hand ships from traditional suppliers also has the advantage that the existing infrastructure is adequate to maintain and operate them and crews require a minimum of retraining. Whenever more recent deletions become available, they are the focus for fierce competition. However, it may be that fewer such vessels will be available in future. Navies in the industrialized world, facing increasing resource constraints themselves, may place greater emphasis on extension programmes and refits designed to stretch the service life of vessels. As a result, ships may not finally appear on the disposal list until they are very old and no longer of interest.

In 1986 and 1987, Chile and Pakistan competed to buy two County class destroyers from the United Kingdom, *HMS Glamorgan* and *HMS Fife*, both of which were sold to Chile in 1987. In 1986, Chile also announced an interest in two Leander class frigates, *HMS Diomede* and *HMS Apollo*, commissioned in 1970 and 1971 and due to be struck from the Royal Navy in 1988. By September 1987, Pakistan and Indonesia had also expressed interest in these ships, described by a Ministry of Defence official as 'the most sought after ships on the Royal Navy disposal list'.[42] They were sold to Pakistan in 1988. Pakistan and Chile are clear examples of traditional regional naval powers facing the twin problems of modernization and severe financial constraint. Prior to the abortive attempt to purchase second hand the County class destroyer *HMS Fife* in the summer of 1987, Pakistan had signed a letter of intent in 1985 with the UK for the construction of two new Amazon class frigates in the United Kingdom and anticipated follow-on licensed production in Karachi.[43] No contract ensued. In 1987 and 1988 negotiations over the purchase of three Type 23 frigates, again from the UK, have stalled over the question of finance. This is in spite of the fact that the ships would have been transferred minus the more expensive and sophisticated on-board systems and that the British Government was prepared to offer Pakistan a reduced interest loan to finance the sale.[44] Chile has, as noted above, succeeded in maintaining a 'traditional' procurement strategy of buying second-hand ships from major naval powers, and has perhaps found the United Kingdom more responsive to its needs since 1982 because of mutual problems with Argentina.

The case of Pakistan illustrates that in countries where the military has had a central role in national politics, the procurement process can be directly affected by domestic political considerations. In Pakistani decision-making for defence, final authority rests with the Defence Committee of the Cabinet, chaired by the Prime Minister and including the Ministers of Defence, Foreign Affairs, Finance, Interior, States and Frontier Regions, Kashmir Affairs, Information, Communications, Commerce and Industry. However, for most of Pakistan's history the Joint Chiefs of Staff Committee, in theory subordinate to the Defence Committee, has included the Chief of Army Staff, who has also been the head of the Government. Moreover, prior to 1976, the Chief of Army Staff was of higher rank than either of the other Service Chiefs. For long periods the Headquarters of the Air Force and Navy were physically removed from decision-making. The Headquarters of the Air Force were in Peshawar (1947–48), then in Karachi (1948–60), then moved back to Peshawar (1960–83) before being moved to Islamabad. Meanwhile Navy Headquarters were in Karachi until 1974, when they moved to Islamabad. Under this arrangement the Pakistan Navy has found it difficult to argue successfully for an equal share of available resources.

Among the other buyers of second-hand naval vessels, there are wide differences in attitudes to naval procurement. The reduced availability of ships with a reasonable life expectancy from the inventories of major naval powers can leave Third World countries with the choice between new ships or no ships at all and to a degree forces naval planners to examine alternative procurement policies. Peru has adopted a mixed policy of buying very old hulls from major navies, which then undergo a major refit, along with a small number of new escorts. In the case of at least four of the vessels refitted—two Daring class destroyers from the UK and two De Ruyter class cruisers from the Netherlands—the process represented a complete rebuild that took over two years to complete.[45] Peru has combined this process with the purchase and subsequent licensed production of Italian-designed Lupo class frigates that led to the induction of four vessels into the fleet by 1986. Peru also expressed an interest in buying one or more frigates built by Argentina under licence from the FRG, which Argentina may no longer be able to afford.

Importers of naval sub-systems

The group of buyers of naval sub-systems includes all countries with a naval shipbuilding industry other than the superpowers, but the navies of the Western alliances represent the most significant market for sub-systems. In many cases sub-systems are imported to increase the working life of units already in service rather than for new construction. Modernization, mid-life equipment upgrades and refits are all an integral part of naval procurement programmes and have been for many years. However, in building new vessels there is also a requirement for imported sub-systems where equipment of the requisite type is not produced indigenously. To give a few examples, in most of the navies of the Western alliances the Standard missile system from the US is the chosen medium-range surface-to-air missile (SAM).[46] British shipbuilders install imported ship-to-ship missiles (ShShMs) on new construction for the Royal Navy—either the French Exocet or the RGM-84A Harpoon from the United States. The Royal Navy and the Royal Air Force (RAF) also buy French SS-11 and SS-12 air-launched anti-ship missiles (ASMs), although the British Sea Eagle and Sea Skua (shortly to become available in a sea-launched version) predominate. In many navies of the alliances the close-in weapon systems (CIWS) for short-range defence against incoming missiles are provided by the Phalanx or Goalkeeper gatling guns from the US and the Netherlands, respectively (although the Goalkeeper system is itself based around a gatling gun from the United States).

Within the Warsaw Treaty Organization there are no countries other than the Soviet Union currently producing naval guided missiles. Consequently the navies of the GDR, Poland, Bulgaria and Romania have all imported naval anti-aircraft and anti-ship missiles from the Soviet Union. Most members of NATO import all of the shipborne helicopters deployed with their fleets either as finished systems or through licensed production, only France, Italy, the United Kingdom and the United States having domestic production capabilities for naval helicopters. Canada and FR Germany have domestic helicopter industries but make no naval versions. As noted above, shipborne aircraft are a sub-system of a very particular kind. Very few aircraft-carriers carry more than 8–10 fixed-wing aircraft, and only the United States currently builds carriers capable of operating a wide range of tactical aircraft. As can be seen from table 2.4, outside of the superpowers, France and the UK, most aircraft-carriers carry helicopters as their primary aircraft.

Table 2.4. Shipborne aircraft complements of selected aircraft-carriers

Owner	Ship name	Fixed-wing aircraft	No.	Helicopters	No.
Argentina	*Veinticinco de Mayo*	A-4Q Skyhawk	5	SH-3D Sea King	
		S-2E Tracker	13	Alouette III	4
Brazil	*Minas Gerais*	S-2E Tracker	6	SH-3D Sea King	6
				AS-355 Ecureuil	2
				A-332 Super Puma	3
India	*Vikrant*	Sea Harrier	6	Sea King	9
	Viraat	Sea Harrier	8	Sea King	12
Italy	*Giuseppe Garibaldi*			SH-3D Sea King	14
	Vittorio Veneto			AB-212 ASW	12
Spain	*Principe de Asturias*	AV-8B Harrier	8	Sea King	8
				AB-212	8
	Dedalo	Sea Harrier	7	Sea King	10
				AB-212	10

Brazil and Italy have aircraft-carriers which, while designed to accommodate fixed-wing aircraft, do not currently do so, reflecting domestic political decisions not to acquire aircraft rather than any technical restraint.[47] Italy changed this policy at the end of 1988 and, as noted above, will purchase limited numbers of sea-based fixed-wing aircraft. With the exception of the US, the USSR and the United Kingdom, all of the navies which operate carriers depend on imported aircraft. Argentina, Brazil, France and Spain operate aircraft from the United States while Brazil and India operate aircraft from the UK. Aircraft carriers are only one among many classes of ship operating helicopters, and helicopters have not only been bought by a wide range of countries but are being used for a widening range of missions. Traditionally seen as platforms for ASW as well as liaison and replenishment, a growing number of countries are attracted by the idea of helicopters as platforms for anti-surface weapons, for target acquisition, communications relay and as airborne early-warning systems.[48]

These examples show that the trade in these major sub-systems is an important element of the naval arms trade equation. In part these imports by shipbuilding countries reflect the lack of a domestic company or industrial group to turn to for a specific sub-system required. As noted above, the buyer will on occasion use the arms relationship to widen the range of domestic industrial capabilities and make use of local employment. However, where the supplier is a close ally a recipient may be satisfied, if not exactly happy, to buy 'off the shelf'.

Notes and references

1   This statistic is in millions of constant 1985 US dollars and is based on deliveries of systems dedicated to naval missions. These systems are ships, for which delivery is taken to mean date of commission, all anti-ship guided missiles, point defence systems and missile fire-control systems and the following classes of aircraft: naval helicopters, maritime patrol aircraft and all fixed-wing aircraft embarked on ships. This figure excludes ships delivered to coast guard forces and merchant ships, even where these would clearly be taken up from trade in an emergency. Also excluded are patrol craft of 100-tonnes displacement or less unless they are armed with missiles, torpedoes or guns in excess of 100-mm calibre, research vessels, tugs and ice-breakers. It excludes torpedoes and ASW rockets and shore-based fixed-wing aircraft.

2   A systematic presentation of data which include information on the trade in naval systems and the production of vessels in the developing world is offered by Brzoska, M. and Ohlson, T., SIPRI, *Arms Transfers to the Third World 1971–85* (Oxford University Press: Oxford, 1987); and Brzoska, M., and Ohlson, T., SIPRI, *Arms Production in the Third World* (Taylor & Francis: London, 1986). Also of interest is Larson, D. L. and Karkoska, A., 'The expansion of naval forces', *World Armaments and Disarmament: SIPRI Yearbook 1979* (Taylor & Francis: London, 1979), p. 357–74. Michael Morris

undertook much of the work for his book *The Expansion of Third World Navies* while a research fellow at SIPRI.

3   Albrecht, U., 'The Federal Republic of Germany and Italy: new strategies of mid-sized weapons exporters?', *Journal of International Affairs*, summer 1986.

4   English, A. J., 'Latin American navies in 1988', *Naval Forces*, Feb. 1988, p. 116.

5   'In return for debt relief, Peru to build 80 ships for Soviets', *Washington Times*, 13 Jan. 1988.

6   Todd, D., 'Government intervention in naval shipbuilding', *Naval Forces*, Sep. 1988, pp. 42–51.

7   Faltas, S., *Arms Markets and Armament Policy* (Martinus Nijhoff: Dordrecht, 1986), p. 75; *Defence Industries in Spain and Portugal*, Western European Union document 1161, 7 Nov. 1988.

8   *Jane's Fighting Ships 1988–89* (Jane's: Coulsden, 1988), pp. 655-57.

9   Mohrez Mahmoud El Hussini, *Soviet–Egyptian Relations, 1945–85* (Macmillan: London, 1987); Ra'anan, U., *The USSR Arms the Third World* (MIT Press: Cambridge, Mass., 1969).

10 El Hussini (note 9), especially chapter 7.

11 The last of these regulations were waived in 1981. Wagner, J., *West German Naval Policy on the Northern Flank: Determinants and Perspectives 1955–88* (Norwegian Institute for International Affairs: Oslo, 1988), p. 181.

12 For example, the F-122 frigate has been built by five different companies in five different yards; the companies involved were Bremer Vulkan, AG Weser, Blohm and Voss, HDW and Thyssen-Nordseewerke.

13 In 1971 the Federal Security Council had established the guideline that weapons would not be exported to 'areas of tension'. The definition of these areas was the responsibility of the Foreign Office. Cowen, R. H. E., *Defense Procurement in the Federal Republic of Germany* (Westview: Boulder, Colo., 1986), p. 30.

14 'The Type-209 submarine: a success story', *International Defense Review*, Sep. 1988, p. 1189, *Jane's Fighting Ships 1988–89* (Jane's: Coulsden, 1988).

15 *Jane's Fighting Ships 1988–89* (Jane's: Coulsden, 1988), p. 204.

16 Faltas (note 7), p. 62.

17 'The second-hand warship market', *Navy International*, Oct. 1988, pp. 478–86.

18 Nathan, J. A. and Oliver, J. K., *The Future of United States Naval Power* (Indiana University Press: Bloomington, 1979), pp. 139–41.

19 *Jane's Defence Weekly*, 20 Feb. 1988, pp. 288–69, 'What to do with de-commissioned US vessels: a Third World solution', *World Weapons Review*, 23 Mar. 1988, p. 15; 'USN decommissions 16 frigates'; *Jane's Defence Weekly*, 2 July 1988, p. 1354.

20 Cavaiola, L. J., 'Congressional Watch', *Proceedings of the US Naval Institute*, Naval Review Issue 1988, pp. 208–12.

21 Vlahos, M., 'Middle Eastern, North African and South Asian navies', *Proceedings of the US Naval Institute*, Mar. 1989, pp. 148–49.

22 *Asian Security 1988–89* (Brassey's: London, 1988), p. 149; 'Asia: Japan boosts defenses with Aegis', *Defense & Foreign Affairs Weekly*, 4 July 1988, p. 2. For a description of Aegis and a discussion of its capabilities, see the interview with Rear Admiral W. E. Meyer, Aegis project manager from 1970 to 1983, in *Proceedings of the US Naval Institute*, Oct. 1988, p. 102–106.

23 *Jane's Defence Weekly*, 23 July 1988, p. 119.

24   Manning, R., 'Zeroing in on Japan; Congress moves to stop US Navy sale of defence system', *Far Eastern Economic Review*, 7 Jan. 1988, p. 21; Bennett, C. E., 'Let's think twice about Aegis', *Armed Forces Journal International*, Apr. 1988, p. 44. Bennett is the Representative who sponsored the motion, which had no legal or prohibitive force in the United States.

25   At the end of March, Congress banned Toshiba from all US Government contracts for three years: Miura, M., 'Toshiba faces 3-year ban on exports to US market', *Daily Yomiuri*, 2 Apr. 1988.

26   'Transfer of Aegis technology to Japan faces further obstacles', *Defense News*, 13 June 1988, p. 50; 'Carlucci ices Japanese Aegis pending congressional action', *Navy News & Undersea Technology*, 17 July 1988, pp. 1, 3; 'Strings on Aegis sale to Japan may cost billions in future', *Defense News*, 25 July 1988, pp. 4 and 37. In the event, the entire Appropriations Act was defeated by Congress and sent back to the White House for amendment.

27   Chanda, N., 'Billions at stake; US Congress tries to block Aegis sale to Japan', *Far Eastern Economic Review*, 18 Aug. 1988, p. 14; 'Senate clears sale of Aegis to Japanese', *Defense News*, 8 Aug. 1988, p. 11.

28   'The Turkish Navy Frigate Programme', *Naval Forces* special supplement, May 1988.

29   *Asian Defence Journal*, Nov. 1988, p. 128.

30   *Jane's Fighting Ships 1988–89* (Jane's: Coulsden, 1988), p. 529.

31   The replacements for the Crusaders on the new class of nuclear-powered CVs under construction by France are planned to be of local design and manufacture—the Rafale—in spite of the preference in the French Navy for a version of the US F/A-18 Hornet. L'Écotais, Y. D., 'Rafale: les extraits du rapport', *L'Express*, 30 Sep. 1988, p. 11; 'Report raises new debate on Rafale costs', *Jane's NATO Report*, 19 Sep. 1988, p. 5; 'Rafale joint radar request, naval version unveiled', *Jane's Defence Weekly*, 10 Dec. 1988, p. 1440.

32   *Jane's Defence Weekly*, 8 Oct. 1988, p. 862; *Defense & Armament Héracles*, Dec. 1988, p. 65; *Interavia Air Letter*, 17 Feb. 1989, p. 5; *Interavia Air Letter*, 24 Feb. 1989, p. 2, 'Aerospace pursues global cooperation', *Wall Street Journal*, 26–27 May 1989, p. 15.

33   'Italy/Spain: Mediterranean security agreement', *Atlantic News*, 13 July 1988, p. 4; 'Shaping Italy's future defence', *Military Technology*, February 1989, pp. 39–43.

34   *Jane's All the World's Aircraft 1968–69* (Jane's: London, 1969) and *Jane's All the World's Aircraft 1988–89* (Jane's: Coulsden, 1988).

35   *World Helicopter Systems* (Interavia data: Geneva, 1985), p. 2.

36   For a detailed breakdown of helicopter sales, see appendix 2.

37   See, for example, *Report of the Indo-American Task Force on the Indian Ocean: India, the United States and the Indian Ocean* (Carnegie Endowment for International Peace/Institute for Defense Studies and Analyses: Washington, DC, 1985).

38   Chesneau, R. (ed.), *All the World's Fighting Ships 1922–46* (Conway's Maritime Press: London, 1980).

39   SIPRI Arms Trade data base.

40   Todd, D., 'Technology transfer and naval construction', *Naval Forces*, Oct. 1988.

41   *Jane's Fighting Ships 1965–66* (Jane's: Great Missendon, Buckinghamshire, 1965), pp. 342–358.

42   *Jane's Defence Weekly*, 19 Sep. 1987, p. 591.

43 Preston, A., 'Trends in Asian Navies', *Asian Defence Journal*, Aug. 1988, p. 10.
44 Urban, M., 'Arms sales policy may be altered', *The Independent*, 18 April 1987; Hussain Zia, 'Frigates for the Navy', *Defence Journal* (Karachi), pp. 32–34; *Defence*, Oct. 1988, p. 812.
45 For fuller details of the programme undertaken in the 1970s, see Brzoska and Ohlson (note 2), pp. 234–35.
46 Hill, J. R., *Air Defence at Sea* (Ian Allen: London, 1987), pp. 69–70.
47 The Brazilian case is discussed in chapter 6.
48 'Strategies of ASW', *Jane's Defence Weekly*, 5 Sep. 1987; Neale I., 'AEW and Naval Operations', *Naval Forces*, Mar. 1987, pp. 92–97; Beaver, P., 'Naval Aviation: a growing EW market', *Navy International*, July/Aug. 1987, pp. 404–407; Hobbs, D., 'New technology and naval helicopters', *Naval Forces*, Oct. 1988, pp. 18–24.

3. Trends in the naval market

I. The proliferation of new naval technologies

Within the wider naval market, certain types of naval system are considered to be worth more detailed attention, specifically the trade in submarines, anti-ship guided missiles and maritime patrol aircraft. on the one hand, and the trade in maritime surveillance assets and smaller patrol craft, on the other. Both trends are responses to changes in the maritime environment as increasing numbers of states have chosen to exercise their maritime responsibilities. In addition, there have been changes in the jurisdictional rights at sea, notably as a result of the signature of the United Nations Convention on the Law of the Sea in Jamaica in December 1982.[1]

The spread of submarines and anti-ship guided missiles have been highlighted by some of the major naval powers as developments of particular concern. In March 1988, testimony to the House of Representatives Armed Services Committee illustrated the extent to which developments of this kind are making an impact on US naval thinking. While historically there have been few if any Third World contingencies which pose a military (as against economic or political) obstacle to the pursuit of US naval objectives, Rear Admiral Studeman, Director of Naval Intelligence, suggested that this may no longer be the case. He noted:

The non-stop proliferation of weapon systems throughout the Third World is both increasing and complicating the tasks of our Navy . . . Weapons in the hands of hostile, potentially hostile, and non-aligned states cover the spectrum from ordinary patrol boats and sea mines through missiles, submarines and highly sophisticated fighter aircraft. SSMs (surface to surface missiles) and ASMs (air to surface missiles) have increased the scope and intensity of Third World ASUW (anti-surface warfare) against both merchants and warships: the most obvious example is the 'tanker war' in the Persian Gulf. Greater numbers of often more sophisticated submarines will tax our ASW capability to distinguish friend from foe and increase the geographic area in which our ASW forces may be required to operate. Expansion and improvements in air systems will further complicate at-sea air defense.[2]

The high seas beyond narrow coastal strips have traditionally been seen as open for commercial and military use. Studeman's comments reflect a concern that the diffusion of increased naval power in combination with changing maritime jurisdictional rights may threaten the extension of a 'creeping jurisdiction' over increasing sea areas and, in particular, over international straits.[3] The issue of creeping jurisdiction is likely to receive greater attention as a consequence of the decision by Indonesia to close the international straits of Lombok and Sunda briefly in September 1988, without either explanation or advance warning.[4]

Such a fundamental change in the perception of the use of the sea has been linked to the trade in naval weapons and equipment by several authors in the past.[5] In its study of the naval arms race the United Nations working group states: 'As with other conventional weapons, there has been a noticeable increase in the demand for the most modern weapons, including anti-ship guided missiles, the delivery of which has provided relatively small coastal navies with a significant increase in war fighting capabilities'.[6]

Part V of the United Nations Convention on the Law of the Sea gave coastal countries rights to exploit an exclusive economic zone to a limit of 200 nautical miles, measured from the baseline of their coastal sea, and places on them a responsibility for the management of this sea space.[7] The LOS Convention has not entered into force because the depositary (the United Nations Secretary-General) has not received the 60 signatures needed for ratification. The legal status of the LOS Convention is ambiguous, and this ambiguity has in itself been a major contributory factor in the spread of maritime surveillance and patrol systems. What is clear is that more and more countries have an interest in gaining a more accurate idea about what is happening in the sea and airspace beyond their immediate territorial waters.

Submarines

All the principal naval powers, including the larger Third World navies, have submarine fleets, and there is a considerable literature on the possible proliferation of submarine fleets.[8] However, in spite of the theoretical attractiveness of submarine forces for smaller navies, their spread has not been rapid. Table 3.1 lists international transfers of submarines during the period 1946–88.

Table 3.1. Deliveries of submarines 1946–88[a]

Buyer	Seller	Designation	Duration of programme	Number delivered
Albania	USSR	Whiskey class	1958–60	2
Algeria	USSR	Romeo class	1982–83	2
Algeria	USSR	Kilo class	1986–87	2
Argentina	USA	Balao class	1960	2
Argentina	FR Germany	Type 209/1	1969–74	2
Argentina	USA	Balao class	1971	2
Argentina	FR Germany	Type TR-1700	1977–86	2
Australia	UK	Oberon class	1963–69	4
Australia	UK	Oberon class	1971–78	2
Brazil	USA	Gato class	1957	2
Brazil	UK	Oberon class	1969–77	3
Brazil	USA	Guppy-2 class	1972–73	7
Brazil	UK	Oberon class	1972–77	1
Brazil	FR Germany	Type 209/3	1982–88	1
Bulgaria	USSR	MV class	1950–52	3
Bulgaria	USSR	Whiskey class	1958	2
Bulgaria	USSR	Romeo class	1972–73	2
Bulgaria	USSR	Romeo class	1984–85	1
Bulgaria	USSR	Romeo class	1985–86	1
Canada	USA	Balao class	1960–61	1
Canada	UK	Oberon class	1962–68	3
Canada	USA	Tench class	1967–68	1
Chile	USA	Balao class	1961	2
Chile	UK	Oberon class	1972–76	2
Chile	FR Germany	Type 209/3	1980–84	2
China	USSR	S class	1954–55	8
China	USSR	Whiskey class	1954–56	2
China	USSR	Whiskey class	1954–64	14
China	USSR	Romeo class	1959–61	4
China	USSR	Romeo class	1959–70	67
Colombia	FR Germany	Type 209/1	1970–75	2

Buyer	Seller	Designation	Duration of programme	Number delivered
Cuba	USSR	Whiskey class	1978–79	1
Cuba	USSR	Foxtrot class	1978–80	2
Cuba	USSR	Foxtrot class	1982–84	1
Denmark	FR Germany	Narvhalen class	1965–70	2
Ecuador	FR Germany	Type 209/2	1974–78	2
Egypt	USSR	MV class	1956–57	1
Egypt	USSR	Whiskey class	1957–58	7
Egypt	USSR	Whiskey class	1962	1
Egypt	USSR	Romeo class	1966–69	6
Egypt	USSR	Whiskey class	1971–72	2
Egypt	China	Romeo class	1980–82	2
Egypt	China	Romeo class	1982–84	2
Egypt	China	Romeo class	1984–86	2
Greece	UK	U class	1943–46	6
Greece	USA	Gato class	1957–58	2
Greece	FR Germany	Type 209/1	1967–72	4
Greece	USA	Balao class	1964–72	2
Greece	USA	Tench class	1973	1
Greece	FR Germany	Type 209/1	1975–80	4
Israel	UK	S class	1956–60	2
Israel	UK	T class	1964–68	3
Israel	UK	Type 206	1972–77	3
India	USSR	Foxtrot class	1967–74	8
India	FR Germany	Type 1500	1981–87	2
India	USSR	Kilo class	1984–	4
Indonesia	Poland	Whiskey class	1958–59	2
Indonesia	USSR	Whiskey class	1958–62	12
Indonesia	FR Germany	Type 209/2	1977–81	2
Italy	USA	Gato class	1951–55	2
Italy	USA	Balao class	1960–72	5
Italy	USA	Tang class	1973–74	2
Japan	USA	Gato class	1954–55	1
Korea N.	USSR	Whiskey class	1960–61	4
Korea N.	China	Romeo class	1973–75	7
Korea N.	China	Romeo class	1973–82	8

Deliveries of submarines 1946–88 continued

Buyer	Seller	Designation	Duration of programme	Number delivered
Libya	USSR	Foxtrot class	1975–78	3
Libya	USSR	Foxtrot class	1978–83	3
Norway	FR Germany	Type XXIII	1950	1
Norway	FR Germany	Type VIIC	1950	3
Norway	FR Germany	Type 207	1959–67	14
Pakistan	USA	Tench class	1963–64	1
Pakistan	France	Daphne class	1967–70	3
Pakistan	Portugal	Daphne class	1975	1
Pakistan	France	Agosta class	1978–80	2
Peru	USA	Abtao class	1952–57	4
Peru	FR Germany	Type 209/1	1970–75	2
Peru	USA	Guppy-2 class	1974–75	2
Peru	FR Germany	Type 209/1	1977–83	4
Poland	USSR	MV class	1953–55	6
Poland	USSR	Whiskey class	1961–69	4
Poland	USSR	Foxtrot class	1987	1
Poland	USSR	Kilo class	1984–	2
Portugal	France	Daphne class	1965–69	3
Romania	USSR	Kilo class	1986	1
S. Africa	France	Daphne class	1968–71	3
Spain	USA	Balao class	1959–74	5
Spain	USA	Guppy-2 class	1971–74	2
Spain	France	S-70 class	1975–86	4
Syria	USSR	Romeo class	1984–85	2
Syria	USSR	Whiskey class	1984–85	1
Taiwan	USA	Guppy-2 class	1972–73	2
Taiwan	Netherlands	Zwaardvis class	1981–88	2
Turkey	USA	Balao class	1948–60	11
Turkey	USA	Balao class	1970–73	10
Turkey	FR Germany	Type 209/0	1974–78	3
Turkey	FR Germany	Type 209/1	1974–	3
Turkey	USA	Guppy-2 class	1977–79	2
Turkey	USA	Tang class	1979–80	1
Turkey	USA	Tang class	1983	1

Buyer	Seller	Designation	Duration of programme	Number delivered
Turkey	USA	Tang class	1979–80	1
Turkey	USA	Tang class	1983	1
Venezuela	USA	Balao class	1960	1
Venezuela	FR Germany	Type 209/2	1971–77	2
Venezuela	USA	Guppy-2 class	1973	1

[a] This table does not record negotiations or undelivered vessels.

The obstacles to operating a submarine fleet are considerable for a country which does not possess a relatively well-developed naval infrastructure. The operating and repair facilities needed for a submarine force are more complex than those required for a surface fleet. In addition to the normal shipyard equipment it is necessary to have equipment for battery recharging and supplying oxygen to the onboard life support systems. Another complicated area is communication with other naval vessels or with land bases. The preferred system for most navies, a very low frequency (VLF) communication system, involves the construction of a shore-based facility costing $25–50 million.[9] For these reasons, the countries which find it easiest to develop submarine forces are those which are building on an existing naval infrastructure.

Since 1971, Algeria, Colombia, Cuba, Ecuador, Libya, Romania, South Korea, Syria and Taiwan have added submarines to their fleets for the first time.[10] Bangladesh cancelled a 1983 order for Chinese submarines. During the late 1970s, Iran contracted to buy West German Type 209 submarines and began making payments for lead items. However, in 1979 the new Iranian government cancelled both this deal and negotiations for Tang class submarines from the United States. There has been speculation that Iran has bought small submarines either from North Korea or Yugoslavia, but there is no confirmation of either report.[11]

It is clear that a significant number of coastal states would like to add submarines to their naval inventories, although it is very unlikely that any will develop large submarine fleets. The marketing director of Vickers Shipbuilding and Engineering Ltd (VSEL)—the principal submarine constructor in the UK—has predicted a market of around 20 vessels among developing countries for his company in the next decade, and the company has developed the Piranha (a submarine of 130–150 tonnes), with this market in mind. In 1987–88 the highest profile prospective client for such a light 'coastal' submarine has been Saudi Arabia, where a final decision over whether to proceed with the procurement of coastal submarines was expected for over a year. The large arms agreement between the UK and Saudi Arabia in July 1988 did not include submarines, and the overall cost of this programme led to the indefinite postponement of an order for submarines.[12]

Malaysia, Nigeria and Thailand are all countries actively contemplating the establishment of a submarine fleet. However, there are good reasons—chiefly technical and financial—to doubt that these three will proceed quickly with ambitious plans. The Commander-in-Chief of the Thai Navy has said that the Government of Thailand will have to choose between buying submarines and other naval equipment. A long-term intention to proceed with the purchase of submarines was part of the 1985 Thai Statement of Defence Estimates, after which Kockums of Sweden and HDW of FR Germany were named as possible suppliers.[13]

Recent statements by the Malaysian Chief of the Navy stating that submarine purchases were planned have been received with some scepticism, given that they have been made every year since 1978.[14] However, Malaysian naval officers have been trained at a school for submariners in FR Germany, suggesting that interest is alive, with actual procurement a long-term objective.[15] In early 1988, Malaysia and the UK came close to agreement on the sale of two second-hand Oberon class submarines to Malaysia, a transfer which had been under negotiation for several years. Negotiations became deadlocked over Malaysian demands for soft financial terms. In September 1988, Malaysia and the UK signed an agreement for the transfer of a range of equipment including one refurbished Oberon class submarine, but whether or not the transfer will ever take place remains unclear.[16]

The attraction of submarines for developing navies stems from their relative invulnerability. Any country which possesses an ocean-going submarine force can threaten retribution against an aggressor, at a place and time of its own choosing. This point was illustrated in the Falklands/Malvinas War between Argentina and the United Kingdom. The UK was able to neutralize the Argentine fleet with the deployment of one nuclear-powered submarine. The Royal Navy submarine *HMS Conqueror* was able to attack and sink the Argentine ship *General Belgrano* without being detected, but even the potential threat from much older and far less capable Argentine submarines created serious problems for the British Task Force. At least one Argentine submarine moved into positions where an attack on elements of the Task Force was possible on at least two occasions, and some reports suggest that attacks failed only because of technical problems with torpedoes. The concern within the Task Force was such that British ships used large amounts of anti-submarine ordnance during the campaign, complicating the war's logistic element.[17]

In spite of the apparent advantages of submarine forces, there remain very few countries capable of operating large numbers of submarines effectively. There have been only two occasions when torpedoes have been used successfully in action since 1945. The Pakistan Navy sank an Indian Blackwood class frigate, *INS Khukri,* in 1971 and the Royal Navy sank the *General Belgrano* in 1982. The problems noted above for countries which would like to buy submarine forces reflect the fact that they impose a high cost in maintenance and manpower terms and demand an extensive infrastructure of land-, sea- and air-based support systems. Assuming that crews are competent (which demands training and exercising, which in turn boost running costs) submarines still find it difficult to find, follow, target and attack surface ships which can adequately defend themselves.[18] Other kinds of submarine campaign—against merchant shipping for example—have been prevented by political rather than military considerations. One author has attributed the limits to the flexible use of naval forces to the fact that naval planners and politicians are 'apt to be dangerously alarmed by eruptions of violence in the sea-lanes of international commerce'.[19] Whether this constraint has survived the war fought by Iraq and Iran against merchant shipping in the Persian Gulf between 1984 and 1988 remains to be seen.

Naval air forces

For many countries naval strategy has come to include a significant maritime air component. Land-based aircraft have a role in all of the major tasks undertaken by navies, but it is not the development of maritime air forces as such which offers a challenge to the naval forces of major powers, since most coastal states have had air forces since soon after their independence. It is more the combination of accurate long-range air-to-surface missiles together with decisions to train air forces specifically for anti-shipping attacks. Anti-ship missiles can be launched from aircraft, surface ships or coastal batteries and the technology required for the submarine launch of such missiles is beginning to spread to such Third World countries as Egypt. The air-launched versions of these missiles have been sold in small quantities compared with the ship-launched versions. However, in recent years events have focused attention on the capabilities of land-based maritime strike aircraft armed with anti-ship guided missiles, rather than ship-launched versions of the same missiles.

The loss of six British ships, including four major surface combatants, during the Falklands/Malvinas War and the attack on the frigate *USS Stark* in the Persian Gulf on 17 May 1987 have between them generated a substantial literature hinting or openly claiming that surface ships cannot be defended against air attack. This literature has generated in turn a considerable quantity of articles rebutting the idea that surface ship vulnerability is either a new or insoluble problem.[20]

In the context of the attacks on shipping by Iranian and especially by Iraqi aircraft during the period between April 1984 and the cease-fire of August 1988, air-launched missiles were identified by British merchant seamen as the primary threat to shipping because of the difficulty of defence. Whereas it was possible to find means of reducing the threat of attack with air-launched bombs by sailing at night or following routes beyond the range of possible threats, the combination of long-range and effective guidance of current-generation missiles made it very difficult for merchant shipping to mount any defence.[21]

For countries which lack the resources to develop surface fleets, the procurement of maritime strike aircraft may be seen as a cost-effective option to enhance coastal defence. The tasks which have been suggested for such air forces include attacks against naval support vessels accompanying enemy ships, attacks on the ships themselves and the laying of mines in areas where ships would be more vulnerable.[22] Existing airfields and air forces might reduce the need to invest in new construction, while the requisite facilities and equipment for training personnel might also serve a dual purpose in training air and naval air forces. Moreover, the capital cost of the aircraft would be less than the investment required to build vessels able to defend themselves against submarine and air attacks in performing the same missions.

However, table 3.2 illustrates that the group of countries whose air forces have bought anti-ship guided missiles is not very large. All of the major naval powers manufacture such missiles. In addition, Israel produces an air-launched version of the Gabriel missile. Air launched anti-ship missiles have been sold within the major Western alliances to Australia, Japan, Italy, the Netherlands, Turkey, the UK and the USA. There is no evidence that any Warsaw Treaty Organization country other than the Soviet Union has any naval versions of air-

Table 3.2. Recent air-launched anti-ship guided missile deliveries[a]

Buyer	Seller	Designation	Years of delivery	Number delivered
Argentina	France	AM-39 Exocet	1979–86	182
Australia	USA	AGM-84A Harpoon	1976–86	60
Bahrain	UK	Sea Skua	1985–87	24
Brazil	UK	Sea Skua	1985–87	32
Egypt	France	AM-39 Exocet	1982–83	40
FR Germany	UK	Sea Skua	1986–88	50
Greece	Norway	Penguin	1976–81	100
India	UK	Sea Eagle	1983–88	156
India	UK	Sea Skua	1985–88	36
Indonesia	France	AM-39 Exocet	1985–86	10
Iran	USA	AGM-84A Harpoon	1972–75	72
Iraq	China	Hai Ying-2	1987	72
Iraq	France	AM-39 Exocet	1978–88	300
Iraq	France	AS-30 L	1985–88	180
Italy	FRG	Kormoran-1	1980–88	82
Japan	USA	AGM-84A Harpoon	1980–88	100
Kuwait	France	AM-39 Exocet	1983–86	24
Netherlands	USA	AGM-84A Harpoon	1978–84	38
Pakistan	France	AM-39 Exocet	1974–83	72
Peru	France	AM-39 Exocet	1982–87	24
Qatar	France	AM-39 Exocet	1983–84	20
S. Africa	France	AM-39 Exocet	1976–80	30
Saudi Arabia	France	AS-15TT	1980–86	220
Saudi Arabia	UK	Sea Eagle	1986–88	200
Saudi Arabia	USA	AGM-84A Harpoon	1986–88	20
Singapore	USA	AGM-84A Harpoon	1985–88	30
Thailand	USA	AGM-84A Harpoon	1987–88	6
Turkey	UK	Sea Skua	1984	36
UAE	France	AM-39 Exocet	1982–84	24
UK	USA	AGM-84A Harpoon	1982	40

[a] This table does not record negotiations or undelivered missiles.

launched missiles. Beyond these countries there have been only 18 customers for missiles of this kind. However, it is notable that the list of buyers contains all of the countries that have coasts adjoining the Persian Gulf, except Oman. Moreover, not only have Gulf countries bought air-launched anti-ship missiles, they have bought them in very significant numbers.[23] In the specific context provided by the Persian Gulf, this has been the root of a great deal of the concern about these missiles.

The performance characteristics of air-launched anti-ship missiles has been less important for their effectiveness than the circumstances of their use. In naval engagements between the US Navy and Iranian forces in the Persian Gulf in 1988, for example, US forces found it easy to defeat by passive means missiles fired at them. US naval forces were ready for an attack of this kind and, because the nature and origin of the threat were known, they could mount an effective defence. The circumstances of the attack on the *USS Stark* were very different in that the attack came from an unexpected direction and was mounted by an aircraft not believed to be hostile. According to the Committee on Armed Services, examining the Iraqi attack on the *USS Stark*, this was more important than questions of hardware. The Committee concluded that: 'There is no evidence that equipment on board the *Stark* failed to work as designed . . . The weapon systems aboard the *Stark* were adequate to the threat it actually faced. There were blind spots where particular weapons could not be brought into action, but that is normal and is a reason why ships maneuver in a hostile environment'.[24]

British naval forces in the Falklands/Malvinas War faced a different kind of problem in combating Argentine aircraft. The post-war inquiries into the conduct of operations have been unanimous in pointing to the shortage of airborne early warning available to British forces once they were beyond the range of UK-based aircraft. The range of shipborne radar is determined by the horizon, and the speed of incoming missiles requires the greatest possible reaction time. The elevation of the radar by putting it on an airborne platform increases the range of surveillance and hence the available time in which to mount a response.[25]

The importance of these aircraft with powerful radars and electronics as a component of a successful anti-missile defence has been acknowledged in a wide range of publications. However, the importance of the same types of equipment to offensive operations has not been acknowledged so widely. The possession of anti-ship missiles is not sufficient in itself to threaten the naval forces of major powers except in some specific circumstances—such as those pertaining in the enclosed waters of the Persian Gulf, where forces are of necessity operating in close proximity to one another. Elsewhere, coastal states seeking to attack the naval forces of major powers face problems of target acquisition unless their aircraft fly so close to the target that they become vulnerable to the air defences of the forces that they are attacking. Aircraft capable of providing the requisite targeting information have not spread beyond the armed forces of major powers. Equally there have been few transfers of systems capable of co-ordinating attacks. Consequently, the understandable focus on events as dramatic as the Falklands/Malvinas War and the attack on the *USS Stark* may well have exaggerated the overall extent to which the gap between the naval capabilities of the major powers and those of naval forces in developing countries has closed.

This is not to argue that this kind of missile proliferation is unimportant, but simply to point out that its importance does not stem from specific technical characteristics of the weapons themselves. Rather, it is derived from the specific context of the Persian Gulf. Naval forces have been central to policies aimed at assuring access to the oil on which the industrial economies depend. While oil production by new suppliers—such as Mexico, Norway, the United Kingdom and Venezuela—reduced this dependency by 1986, over the long term the place of the Persian Gulf littoral countries as oil exporters is likely to reassert itself.[26] Whether any of these countries singly or in combination can keep the Straits of Hormuz closed (with the kind of systems discussed here in combination with other military means such as mining and the use of coastal missile batteries), in the face of a response by the major naval powers, is an open question. The fact that there is a question to answer is evidence of a change in the global political environment where major powers used to exercise greater authority. However, testing the implications of this new environment by military means could not meet any of the interests of the affected parties, which reflects that issues such as freedom of navigation are increasingly questions for political dialogue within and outside the region rather than security problems as such.

II. The impact of maritime law

Maritime patrol, low-level naval contingencies and constabulary operations

As an increasing number of young states have begun in the 1980s to identify, define and exercise new maritime responsibilities in response to changes in the jurisdictional rights at sea, choices concerning what to buy, how much and from whom have become more complex. As noted above, Part V of the Law of the Sea Convention gave coastal countries rights to exploit an exclusive economic zone. The fact that the Law of the Sea Convention has not entered into force has not prevented a large number of countries from claiming such a zone, as defined in the Convention. Arguably, these claims have no status in international law, but they have acquired a form of legitimacy through usage and statements by major power non-signatories—especially the United States—that they will observe and respect portions of the document. This ambiguity has been a major factor in the spread of maritime surveillance and patrol systems. Signatories of the Law of the Sea Convention do not have sovereignty over their exclusive economic zone, but they have been awarded jurisdiction over certain specific activities. Other parties have rights of transit and overflight, and the Law of the Sea Convention does not countermand or override other international laws. The language of the Convention is intentionally ambiguous in delimiting the rights and obligations of coastal states, but it is clear that they have a responsibility for overseeing the activities in their exclusive economic zone. Fishing vessels as much as any other have rights of transit through the EEZ. A coastal state cannot legally prevent the passage of a fishing vessel or fleet, but such states have an interest in monitoring the activities of boats in the waters of their exclusive economic zone. Because of the lack of clarity over the legal position of the Law of the Sea Convention, whether a coastal state has the right to do more than monitor the activities of such vessels is open to question. Whatever the law says about their actions, certain countries have decided that they will take action against foreign fishing vessels within their EEZ, and they have invested in systems that will allow them to do so more effectively.

Demarcating national boundaries has traditionally been a function for land forces. However, the introduction of the EEZ has made the demarcation of a new form of maritime boundary an important issue. In places where states share congested coastlines or border on semi-enclosed stretches of sea there is a problem of allocating sea area. Here questions of surveillance and patrol quickly become meshed with the possibility of naval conflict. The process of maritime reconnaissance and patrol will be subsumed beneath other elements of maritime security. However, although disputes over maritime jurisdiction in contested sea areas are potential sources of conflict, they need not be so.

There are examples where questions of maritime boundaries have been the catalyst for regional co-operation. Moreover, this has on occasion led to limited sacrifices of sovereignty to supra-national bodies, notably by Malaysia and Indonesia in the context of The Association of South-East Asian Nations (ASEAN). For example, the ASEAN Committee on Petroleum and that on Marine Pollution have been influential in making decisions in several areas. In 1971, Singapore, Malaysia and Indonesia agreed on a tripartite statement on the Straits of Malacca and Singapore based on transit passage provisions then emerging in the Law of the Sea Convention which allowed free passage of ships, although the governments of Malaysia and Indonesia disagreed with the proposition that the Straits of Malacca were international waters.[27] Malaysia, Indonesia and Singapore have also agreed on a scheme for the separation of ship traffic and imposed a minimum under-keel clearance for vessels in the Straits of Malacca although the countries still formally disagree on the legal status of the waters. This is because they acknowledge that ship collisions and cargo spillages damage all their interests.

The growing demand for equipment which allows combined operations by maritime patrol aircraft and patrol vessels illustrates the organic link between arms procurement and local perceptions of security.

Small countries—even some micro-states—invest large amounts of time and money in the acquisition of maritime surveillance and patrol capabilities. Of the micro-states Belize, Cyprus, Guyana, Jamaica and the Seychelles operate Britten Norman Maritime Defender aircraft equipped with Bendix RDR-1400 radar. The Marshall Islands operate two Searchmaster aircraft made by GAF (the Australian Government Aircraft Factories) with the same Bendix radar. Papua New Guinea bought six Norman 22B Missionmaster/Searchmaster aircraft between 1981 and 1986 which it uses in the maritime surveillance role. Gabon has bought a Brazilian EMB-111 with a US-supplied search radar and Mauritania has a Piper Cheyenne aircraft also equipped with the Bendix RDR-1400.[28]

The fact that countries with such limited resources choose to allocate them in this way suggests that for them security has a far more complex definition than issues of inter-state war and peace. The possibility of a major conflict is a contingency of which planners have to be aware, especially where there is a dispute over maritime jurisdiction. However, more routine but equally important functions have to be performed. Navies have a role in counter-insurgency and protection from subversion, in maintaining maritime law and order and in safeguarding access to maritime economic assets. The first step in looking after any maritime interests has to be visible and demonstrable surveillance and patrol of existing national sea space. For most countries, therefore, the starting-point in addressing maritime security needs is the surveillance, data processing and communications without which it is impossible to know what is happening in a large ocean area. Moreover, this information has to be supported by national means for practical and political reasons. In extreme circumstances such as a crisis or war, intelligence may be available from a major power as a function of alliance membership or because it suits the purpose of that power to disseminate the information. This is not likely to be the case where contingencies are of a lower order of magnitude, occur as a matter of routine or are purely domestic in their implications.

As indicated in appendix 2, a wide range of countries have invested in capabilities to undertake some form of maritime surveillance from the air. Column five indicates that fixed-wing aircraft designed for maritime patrol can cover considerable areas of sea surface during their sorties. Airborne sensors allow a large sea area to be covered in a relatively short period of time, particularly if there is a network of coastal air bases that allows more flexible patrol routes to be covered. Offshore assets are in specific locations and other legal responsibilities such as the maintenance of law and order are also confined by geography to a degree. Piracy must follow shipping trade routes, coastal geography may limit the places where insurgents or smugglers can put ashore and so on. Consequently it may be that even small countries can make a considerable national effort at effective surveillance if the missions planned are specific, rather than speculative patrolling. However, for most countries a comprehensive knowledge of all activities within the offshore estate is impossible; there is simply too much sea to survey. Looking at the data on maritime geography in appendix 2, the vast size of the area of sea space over which island or archipelago countries claim some jurisdiction can be seen. Even accepting the surveillance capabilities as presented in column five of appendix 2, it would be impossible for countries such as Australia or Indonesia to maintain a comprehensive picture of activities in their respective EEZs. In fact, the capabilities of the aircraft presented tend towards overstatement because of the methodology chosen.

The search capability per day is based on the number of sorties an aircraft might make per day, but in countries with limited ground support facilities and aircrew it may not be possible to turn around aircraft quickly. Moreover, few countries have the air base infrastructure or the financial resources to sustain costly patrols. As an indication of the costs, the US Navy and Air Force flew over 6000 hours of patrols in the first six months of 1986 in support of drug law enforcement in the Caribbean at a cost of some $7000 per hour.[29] Even after such a colossal investment, one estimate suggests that only 10 per cent of surface traffic was identified successfully.[30]

There are obvious limits to the effectiveness of aircraft in low-level security operations. In particular they face the problem of interdiction. From the air there is a stark choice once a suspicious vessel has been detected below. Either the vessel can be tailed in the hope of pursuing enquiries once it puts ashore, or it can be attacked without further investigation. It is easy to think of situations in which aircraft would be completely incapable of carrying out missions without the support of surface vessels—for example, where the purpose of a ship under surveillance is unknown. Surface vessels offer many more possibilities, ranging from shadowing the vessel at a distance to engagement and boarding, and the implications of managing the offshore estate therefore extend beyond the trade in reconnaissance aircraft to the sale of patrol vessels capable of meeting this specific maritime role. There is no absolute demarcation between the various levels of maritime security policy, and clearly on occasion larger naval forces can be called on to fulfil this role. In some cases, countries have established special forces to meet specific needs, although these forces normally sustain institutional links to the defence establishment so as to be available in a crisis.

Egypt, Indonesia, Morocco, Peru, the Philippines, Saudi Arabia, South Korea and Taiwan have in the past five years all bought ships for coast-guard-type forces that are armed and over 100 tonnes in displacement. Peru and the Philippines are countries that have embarked on coast guard programmes at a time when their navies are finding it difficult to obtain funding for equipment of any type.[31] Patrol vessels are required to conduct visible operations in order to be effective, and this means that several ships are needed and that these must remain on station for extended periods. The need for a rugged, cheap vessel excludes more heavily armed craft. The addition of advanced cruise missiles or high-calibre guns with their associated target acquisition/fire control systems raises the capital cost of ships dramatically. It also raises running costs. Not only are repair, maintenance and resupply more complicated, but the effective use of these weapons also requires specialist crewmen who may be poorly trained for other duties. Apart from the financial implications of mounting advanced weapon systems, together with their related ordnance they take up space, necessarily reducing the carrying capacity for other equipment. Effective day-to-day management of the offshore estate and an investment in missile-armed corvettes may be alternatives for countries with limited financial resources.

Notes and references

1   The text of the Law of the Sea (LOS) Convention is reproduced in *International Legal Materials*, no. 21, Nov. 1982, pp. 1261–354.

2   Statement of Rear Admiral William O. Studeman, Director of Naval Intelligence US Navy, before the Seapower and Strategic and Critical Materials Subcommittee of the House Armed Services Committee on Intelligence Issues, 1 Mar. 1988, pp. 53–54.

3   Darman, R. G., 'The law of the sea: rethinking U.S. interests', *Foreign Affairs*, vol. 56 (Jan. 1978); Larson, D. L., 'Naval weaponry and the law of the sea', *Ocean Development and International Law*, vol. 18, no. 2, 1987; United Nations, *A Quiet Revolution: The United Nations Convention on the Law of the Sea* (United Nations Publication: New York, 1984); US House of Representatives, Committee on Foreign Affairs, Hearings, 17 June, 17 Aug., 16 Sep. 1982; Boothby, D., 'Maritime change in developing countries', paper for Conference on Naval Forces and Arms Control, SIPRI, 1–3 Oct. 1987.

4   Chanda, N. and Holloway, N., 'Troubled waters: Indonesia's neighbours consider response to closure of Straits', *Far Eastern Economic Review*, 10 Nov. 1988, p. 18; Leifer, M., 'Indonesia waives the rules', *Far Eastern Economic Review*, 5 Jan. 1989, p. 17.

5   Larson and Boothby (note 3). See also Morris, M., *Expansion of Third World Navies* (Macmillan: London, 1987); and the United Nations study series on

disarmament, no. 16, *The Naval Arms Race*, sections 74/80, United Nations publication A/40/535, New York, 1986.

6   *The Naval Arms Race* (note 5), section 79, p. 21. Another commentator identifies a 'burgeoning' of countries with maritime forces and the prospect that 'where there's one flower now soon there may be a garden', Moore, J. E., *Jane's Fighting Ships 1987–88* (Jane's: London, 1987), p. 125.

7   Part V of the Law of the Sea Convention is reproduced in Westing A. H., SIPRI, *Global Resources and International Conflict* (Oxford University Press: Oxford, 1986), pp. 233–60.

8   See for example Preston, A., 'Establishing a submarine force', *Asian Defence Journal*, Jan. 1988, pp. 44–50; Lewis Young, P., 'Modern conventional submarines: present trends and future developments in the Asian–Pacific region', *Asian Defence Journal*, Oct. 1987, pp. 62–78; 'The submarine: export success', *Navy International*, May 1988, p. 224.

9   Preston, A., 'Establishing a submarine force', *Asian Defence Journal*, Jan. 1988, p. 48.

10 Brzoska, M. and Ohlson, T., *Arms Transfers to the Third World 1971–85*, SIPRI (Oxford University Press: Oxford, 1987).

11 *SIPRI Yearbook 1988: World Armaments and Disarmament* (Oxford University Press: Oxford, 1988), p. 232. It is not clear at the time of writing whether the submarines ordered from FR Germany will be delivered.

12 *Defence*, Aug. 1988, p. 548.

13 *Asian Defence Journal*, Aug. 1987, p. 100.

14 'Malaysia's navy plans', *Jane's Defence Weekly*, 5 Mar. 1988, p. 387.

15 'Malaysia throws doubt on UK Oberon deal', *Jane's Defence Weekly*, 6 Aug. 1988, pp. 200–201.

16 The barriers to the deal have been financial. Discussions continue concerning the level of bilateral trade and British aid to Malaysia to be provided through the Overseas Development Administration; 'Malaysia throws doubt on UK Oberon deal', *Jane's Defence Weekly*, 6 Aug. 1988, p. 200; Eglin, R. and Adams, J., 'Scoring on defence', *Sunday Times*, 2 Oct. 1988, p. D9; Harrison, M., 'UK credit for £1 bn Malaysian arms deal', *The Independent*, 10 Sep. 1988.

17 Scheina, R. L., 'The Malvinas campaign', *Proceedings of the US Naval Institute*, Naval Review Issue 1983, pp. 106–107; Scheina, R. L., 'Where were those Argentine subs?', *Proceedings of the US Naval Institute*, Mar. 1984, p. 1156; Friedman, N., 'The Falklands War: lessons we learned and mislearned', *Orbis*, winter, 1984, p. 896.

18 For a description of defence mechanisms against submarines see Daniel, D. C., *Anti-submarine Warfare and Superpower Strategic Stability* (Macmillan: London, 1986), part I.

19 O'Connell, D. P., 'Limited war at sea since 1945', in M. Howard (ed.), *Restraints on War: Studies in the Limitation of Armed Conflict* (Oxford University Press: Oxford, 1979), p. 126.

20 For a discussion of the two sides of the debate, see Hill, J. R., *Air Defence at Sea* (Ian Allen: London, 1987).

21 Third Special Report from the Defence Committee Session 1986–87, 13 May 1987, *The Protection of British Merchant Shipping in the Persian Gulf* (HMSO: London, 1987), pp. 78–81.

22 For a general discussion of the issues of naval aviation see Sullivan, W. K., 'Now is the time to rethink, redesign and redeploy naval aviation', *Naval War College Review*, Mar.–Apr. 1982, pp. 10–17.

Part II. Case studies

4. Collaborative naval programmes

I. Introduction

The progressive movement away from wholly national industry or self-sufficiency in defence production has reflected the more demanding technological characteristics of recent generations of weapon systems as well as the growing cost of research, development and production. The spiral is characterized by one defence contractor as follows: 'Budget pressures at best result in fewer systems for the same money and at worst in less money and even fewer systems. Fewer systems in turn require higher technology if the quantitative disadvantages are to be overcome. This means that the R&D process takes longer, by which time the threat has probably increased, and so on'.[1]

The changing nature of the military industrial process has also reflected a perception that structural overcapacity for military production represents a major waste of resources. Historically, the ability to increase the volume of military production in times of crisis or war was regarded as a strategic necessity. However, individual weapon systems currently being produced are of a level of sophistication that disallows rapid surges in output. Moreover, were such an increase in output possible, training the manpower to operate these systems and integrating them into the armed forces would be a lengthy process. Equally important have been changes in perception about the practicality of world war (or wars between major powers) in the nuclear age, which have produced a consensus that there is no conceivable benefit to be derived from unlimited war. As one author describes the equation: 'multiply the concept of unrestricted total warfare by the power of the hydrogen bomb to obtain the apocalypse'.[2] In this environment the necessity for preserving surplus productive capacity on strategic grounds has been overtaken by issues that include the economic burden of sustaining the defence sector and the military disadvantages of lack of standardization in equipment.

These factors have contributed to the pressure for increasing collaborative production. The programmes can take different forms. On the one hand, there could be the growth of binational or multinational companies, a move away from the traditional structure of the defence industry which has been characterized by nationally owned contractors. On the other hand, collaboration could take the form of fluid industrial partnerships, that is, alliances between nationally owned companies formed to bid for specific contracts and then dissolving or reviewing their relationship once the contract was either completed or lost. This is not a development confined to naval procurement. Large naval programmes have been isolated as a particularly difficult area for co-operation because of the problem of synchronizing the design requirements and procurement cycles of participating navies. However, the naval market does present examples of co-operative procurement.

II. The political impulse to co-operate

The technical pressures to collaborate on defence programmes and reduce overall production capacities are not exclusively international. In the United States, for example, there has been a considerable rationalization of the defence industrial base that has reduced the number of defence contractors in certain sectors. Whereas in the 1960s there were five companies able to produce nuclear-powered submarines, there are currently only two. However, the pressures to make this process multinational are often political.

The question of NATO burden-sharing has highlighted the fact that the potential impact of co-operative production on the trade in naval systems is a function of political relationships within the alliance as well as technological and economic considerations. This has been shown in particular by the passage in the United States of the amendment to the Fiscal Year 1986 Defense Authorization Bill sponsored by Senators Nunn, Glenn, Roth and Warner. The amendment provides $200 million specifically to improve research and development collaboration among members of NATO and disallows the use of the money for any kind of procurement of equipment.[3] The money was to be divided between the services, and so the US Navy in a sense has had collaborative programmes thrust upon it. Senator Roth explained the need for greater collaborative effort as follows:

America's European allies are unlikely to accede to Congressional requests to build up their conventional forces if such requests are equated with a signal to 'buy American' . . . The United States armed forces cannot expect to procure a new generation of weaponry if it persists in its traditional policy of 'make it and buy it in America'. Similarly, while the current push to 'go European' on the other side of the Atlantic is understandable, its advocates must know that this policy does not guarantee production runs of sufficient lengths to stabilize prices while, in some critical fields, Europe simply lacks the necessary technological expertise . . . NATO members must abrogate their current nationalistic, disparate defense efforts and begin to function like a true alliance.[4]

The US Navy announced that under the terms of the amendment it was interested in the joint development of a new sonar with Canada, the Netherlands and the UK; an Identification Friend or Foe system for ships in collaboration with FR Germany, Norway and the UK and a torpedo programme in conjunction with the UK.[5]

The idea of collaboration is of enormous potential significance for the arms trade. For the members of the NATO alliance, genuine co-development would assist in alleviating the perennial problem of the imbalance in intra-alliance arms trade, which sees the United States win the lion's share of competitive orders. Certain systems previously traded between members of the alliance would be developed and produced jointly. This prospect has been made explicit as a factor promoting the idea of co-development. The possible growth of co-development programmes is also important for its implications in the area of exports of naval weapon systems. As is stressed in chapter 1, the global market for new warships and other types of naval system has been dominated by European countries and the Soviet Union and the export policies of suppliers have been under close national control. The sale of co-developed systems would require a different structure for managing exports beyond the participant countries.

The issue of exports beyond NATO has been an obstacle to the development of trans-Atlantic co-operation because of concerns in European countries, notably France, about regulation. A 1983 survey of the attitudes to co-development among French industry managers and government officials identified this as a crucial consideration. One French official noted: 'We find that if we want to sell something that . . . was involved in a co-operative effort with the US, we have to get a complete OK before we do anything for export sales. We can understand this within limits. But it usually gets to the point where the US wants to have near total control of everything we are doing'.[6]

If it is clear that the impact of co-operative production on the arms trade is potentially very important, it is equally clear that the progress of co-development programmes is closely linked to a complex mesh of economic, military and political questions in an area that has traditionally been a national preserve. A former chairman of the IEPG, Jan van Houwelingen, has expressed the problem thus: 'defence is closely tied to national sovereignty . . . and for many years governments and industries have collaborated to protect their national markets from foreign defence equipment'.[7] Co-development requires a participating country to accept a degree of loss of national control over the procurement process, and therefore presupposes a close political relationship between participants. Reference has already been made to the immense complexity of managing programmes, and the feasibility of co-development requires the existence of an institutional framework that will allow the requisite co-ordination and planning. Van Houwelingen also noted that co-development required participants to 'harmonize national equipment requirements and time scales for the acquisition of defence equipment' since once nations have defined their requirements and industries have developed the prototypes, co-operation becomes more difficult.[8] Therefore, co-development also requires consultation and joint planning by participants from the outset. The possibilities for co-development are restricted to a small number of countries by these criteria, and in the foreseeable future perhaps do not extend beyond NATO.

III. Collaborative naval programmes

Given the definition applied here, naval co-development programmes have a very recent history but there are now several programmes that can be discussed as such.

One such programme which had already led to the construction of 28 vessels by the end of 1988 is the Franco-Dutch-Belgian tripartite programme to build a minehunter, designated the Eridan, Alkmaar and Aster class by the respective navies.[9]

The background to the tripartite minehunter programme was the simultaneous requirement among the three navies for a ship of broadly similar characteristics and level of performance. The programme began in 1977 as a co-development programme but quickly developed a strong French project leadership. The initial requirement was for a total of 45 vessels, 15 for each of the participating navies. However, first Belgium and then France reduced their requirement to 10.[10] The project office was established in Paris and the prototype was developed in France between 1977 and 1981. The design was then released to the Netherlands and Belgium, which caused some annoyance in the Netherlands.[11] One of the consequences of the collaborative approach to this programme has been the development of a number of alternative equipment fits within the same hull. The Dutch producer of the Alkmaar class, the Van der Giessen-de Noord shipyard, offers alternative sonars, communications systems, armament and data processing systems, allowing customers to match these modules with equipment already in their navies.[12]

In the case of naval weapon systems, there are prospects for the future co-development of torpedoes and naval mines. Torpedoes and naval mines were singled out by the United States Congress in 1986 as possible areas for future co-development, and in particular, there are prospects for the merger of studies being initiated in the United States and the United Kingdom for the production of a Surface Ship Torpedo Defence System (SSTDS).[13] Italy and France have agreed the co-development of an anti-submarine missile called Milas based on the French Murène torpedo and the Italian Otomat missile.[14]

In spite of its name, the NATO Sea Sparrow anti-aircraft missile is entirely of US design, part of a family which originated with the air-launched AIM-7, a programme stretching back to 1950.[15] In 1968, the US, Denmark, Italy and Norway signed a Memorandum of Understanding (MOU) to produce the missile but none of them has done so. Canada and Italy have production arrangements. In Canada production is undertaken by the Raytheon company of Canada, and is heavily derivative of the US programme. In Italy, although a licence was signed by Selenia and facilities for Sea Sparrow production exist, no production has taken place and the company makes a competitive system called Aspide.

The production of a NATO Anti-Air Warfare System (NAAWS) for the NFR-90 has become intimately connected with the NATO frigate programme and is discussed in greater detail below. There are currently two alternatives under discussion, both involving consortia of industries from several NATO countries. Other alternatives are expected, but all of those proposed involve the combination of existing national programmes in some way rather than the initiation of a new development programme. Its intimate connection with the progress of the NFR-90 makes it impossible to ignore the NAAWS programme, but it is discussed here only in the context of the NFR-90 and not in depth. This is in part necessitated by the fact that many decisions concerning the status of the project have yet to be taken and it is too close to events for anything but a limited and tentative discussion.

NFR-90

The NFR-90 meets the criteria of co-development outlined, as a brief description of the programme so far makes clear. In 1979, the NATO Naval Armaments Group (NNAG), given the task of defining a NATO Staff Target, created Project Group 27, consisting of the representatives of seven countries—Canada, France, FR Germany, Italy, the Netherlands, the UK and the US. In the case of the NFR-90, the fact that from the outset this was a NATO programme did not prevent French participation. By the end of 1980, this group had concluded that a requirement existed for a conventionally powered ship of roughly 3500 tonnes. However, there was disagreement over the primary role of the vessel. Four members saw the vessel primarily as an ASW frigate, while Canada, the US and FR Germany wanted the vessel primarily for air defence. Therefore the Group concluded that while a common hull design was possible, the design of the ship should be 'modular', allowing each navy to specify its own equipment fit.[16] The NATO Industrial Advisory Group (NIAG) was then given the job of carrying out a feasibility study, identifying the costs and benefits of different ways to proceed with the programme and recommending one.

In February 1981, NIAG Sub-Group 13 began to evaluate the various solutions to the requirement in terms of operational capabilities. Drawing on over 130 companies, they presented a report in October 1982 that discussed 12 possible designs ranging from 2500 tonnes to 4000 tonnes. The study concluded that a ship of 3500 tonnes could offer the flexibility to operate in the very different sea conditions faced by the navies of the participant countries as well as to support the range of different weapon, sensor and logistic systems required. Moreover, a successful co-operative programme might offer savings of about 20 per cent on research and development and a 12 per cent saving on life-cycle operating costs.[17] In 1983 each nation initiated its own assessment of the NIAG recommendation, initially for presentation to the Conference of NATO Armament Directors at a spring 1983 meeting. These national assessments would also be used comparatively later in the programme to see whether collaboration really offered significant advantages over more traditional methods of procurement.[18] At the meeting, Belgium and Norway, which had been members of the NIAG feasibility study together with Spain and the seven original participants, decided that NFR-90 could not meet their national requirements and withdrew from the programme. However, eight countries participated in drafting a Memorandum of Understanding that had to be signed by April 1984 to qualify for membership of the programme. A more detailed feasibility study leading to the acceptance of a NATO Staff Requirement was also set up to identify specifically the systems needed to meet the national requirements of participants. This feasibility study was to be co-ordinated by the Project Management Office (PMO) for the programme, based in Hamburg, and assigned two naval officers from each of the eight participating nations. This group had the job of liaising between the national navies and a private sector joint venture company—the International Ship Study Company—composed of representatives of lead companies, one from each country, nominated by participating countries.[19] This nomination created a problem for the United States in that their national procurement regulations required the government to issue a Request for Proposals (RFP) and select a prime contractor from the bids.[20] For other participants the process of nomination was simpler because there were very few candidates in each country and no legal obligation for a competition.

The feasibility study was completed on 29 October 1985 and delivered to the Project Steering Committee of the NATO Naval Armaments Group. Here it was compared to possible national solutions to the frigate requirement prior to a decision, expected by January 1987, about whether to proceed to a project definition stage. In fact, the Statement of Intent to participate in the next phase was signed on 29 July 1986 by the eight participating nations, although several caveats were inserted by members.[21] The project definition phase was intended to specify common equipment in all of the ships, expected to amount to 50 per cent of the total equipment on board, and to select the suppliers by competitive tender.[22]

The co-development of the NFR-90 has become closely linked in the minds of the potential customers with the development of the on-board equipment, and in particular the air defence system. In January 1988, the British Government made the link explicit with the decision to proceed with the project definition phase only on condition that the timetables for ship production and weapon system fitting were brought into alignment at 'an early stage'.[23] Of these systems, the largest and most complex is the air defence system for the vessel—the NATO Anti-Air Warfare System. This programme is discussed in more detail below.

It is important to note that among the participant nations of the NFR-90 project, none of the major shipbuilding nations has foreclosed the option of a predominantly national solution to its frigate requirements. Spain is not included as a major shipbuilding nation here. The NATO frigate is peripheral to the procurement plans of the US Navy, and the suspicion that the ship could not meet US Navy requirements has been ever present. These reservations emerged most clearly during the discussion of the feasibility study prior to the signature of the MOU in April 1984. Assistant Secretary of State Richard de Lauer, apparently at the behest of the Pentagon legal department, produced important qualifications concerning US involvement in the programme, which were only withdrawn after intervention by George Shultz and Caspar Weinberger.[24] However, the last of the FFG-7 class of frigates is expected to be completed by 1990, and funding for a successor programme, the Future Frigate (FFX) programme, was halted in April 1986. This programme itself envisaged the production of around 50 ships, and there have been suggestions that the NFR-90 could meet this requirement.[25]

Canada, France, the FRG, the UK and Italy all have national programmes under way which will lead to the production of new escort vessels which superficially are of roughly the same design and tonnage as the NFR-90. The programmes are as follows.

In Canada, the first of six City class patrol frigates began production in March 1987 and was launched in January 1988. These ships were not conceived as alternatives to the NFR-90, but were intended and will be equipped for anti-submarine warfare. However, there is no reason why they should not be fitted with a different equipment fit specifically for air defence. The production of these ships is part of a Ship Replacement Programme (SRP) that was the product of a long period of evaluation and discussion stretching back as far as the 1969 Defence Review and the subsequent 1971 Defence White Paper.[26] In May 1972 there was a further Maritime Policy

Review and in 1975 a Defence Structure Review. A formal announcement of intent to build six new ships came in 1978, but an invitation for tenders was not issued until August 1983.[27] By 1982 a Cabinet decision on the size of Canada's requirement for frigates had been taken and was revealed during testimony to the Senate Sub-Committee on National Defence in June 1983. On 18 August 1983, the contract was awarded. The SRP was originally planned to have three phases, each involving the building of six ships. However, a decision to follow the purchase of a second batch of six Halifax class frigates with a further six was cancelled.

In France, funds for a new air defence escort were allocated in the 1978 defence budget, but the first ship was laid down only in September 1982 and launched in February 1985. A second ship was laid down in 1986, but subsequently work and new orders have been suspended pending consideration of the air defence systems on board. The first ship was armed with the Standard SM-1 system from the United States and initially this was to be the system adopted for all of the ships. However, later ships may now be fitted with the Family of Missile Systems (FAMS) discussed below.[28] In April 1988, the French Ministry of Defence also announced the development of a new class of light patrol frigate of roughly 3000 tonnes to be designated the FL-30.[29]

In the Federal Republic of Germany the production of up to eight new air defence escorts is planned, possibly of two different types. The design of four Type 123 frigates may be based on the existing Type 122 Bremen class—itself a relatively modern design based on the Dutch Kortenaer class ship, the first ship having been commissioned in 1982.[30] The Type 123 would be of a higher tonnage but would incorporate a great deal of standardized equipment already being bought for the Type 122 vessels. These vessels will be built as part of the West German Navy's modernization programme regardless of the progress of the NFR-90; the first is to be laid down in 1989. However, included in the planning of the Type 123 is the possible future construction of an additional four Type 124 air defence escorts whose specifications have not been revealed. The successful production of the NFR-90 might meet this requirement.

In Italy, two new air defence escorts are currently being built and are planned to be commissioned in 1991 and 1992. These ships, the Animoso class, are specifically designed for air defence.

In the Netherlands, the first of a new class of frigate, the Karel Doorman class, was laid down in 1985 and launched in April 1988. This class was itself a development of the Kortenaer class, the last of which was completed in 1983.

In the United Kingdom, current shipbuilding plans for surface ships centre on the production of Type 22 and Type 23 frigates, both of which are intended for anti-submarine missions. For air defence, the programme to construct 12 Type 42 destroyers was completed in 1985 with the commissioning of the *HMS York*, and the focus of medium-term plans would be the retrofit of equipment on these vessels. The construction of surface vessels has become a subject of controversy as the influential Parliamentary Defence and Public Accounts Committees have criticized aspects of shipbuilding programmes. In particular criticism has been levelled at the rate of orders for new frigates, the progress of development of the computer systems for command, control and communications and the overall management of the Type 23 programme, production of which has suffered severe cost overruns. In this climate, and given the fall in real terms in British defence expenditure over the medium term, there is little or no money available for the construction of an entirely new vessel which the Royal Navy will not require until the late 1990s.

It is worth noting in passing that many of these 'national' solutions to the requirement addressed by NFR-90 will in themselves involve a substantial foreign input since most are planned to incorporate significant imports of naval weapon systems as they develop. In many of these navies the close-in weapon systems (CIWS) for short-range defence against incoming missiles are provided by the Phalanx or Goalkeeper gatling guns from the US and the Netherlands, respectively. Canada will import the helicopters deployed with its escorts and many other navies use helicopters produced under licence from the United States or France. Of the new series in production in Canada, France, the Federal Republic, Italy and the United Kingdom, only the French ship will not incorporate an air defence system of foreign origin. The Canadian ships will make use of the Sea Sparrow; the West German vessels will also use the Sea Sparrow, and in addition will be armed with Stinger and Rolling Airframe Missiles. Italian air defence escorts will make use of the Standard SM-2 missile. Dutch air defence frigates will be armed with the Sea Sparrow. All of these missiles were either bought directly or developed from designs that were licensed by the United States.

More important for our immediate purpose of discussing co-development are other observations regarding NFR-90. None of the participant countries in the project feel sufficiently confident that the programme will reach successful completion to forgo national alternatives. One of the most important considerations for the future development of the NATO frigate will be the future shipbuilding plans of the US Navy, and in particular the manner in which the US Government will replace 46 Knox class frigates currently in service. If these vessels were to be replaced by the NATO frigate design rather than by a wholly US vessel, it would dramatically change the magnitude of the NFR-90 programme.

The existence of these national programmes in turn tends to reduce the requirement for the NFR-90, since there are a finite number of systems needed by the respective navies. This fact lies behind the often repeated claim that the success of co-development is dependent on a political rather than technical analysis of procurement possibilities.

The impetus behind co-development has in part come from the perception that some programmes are simply so costly that it is not possible for one country to undertake them. However, there seems to be a question whether there is sufficient commonality among the requirements of so many navies to devise a coherent programme at all. The assumption on which this rests—that there is a requirement for escort vessels of over 3000 tonnes in major navies—may be true in the broad sense, but it is clear from the above discussion that the timing of the requirement for the NFR-90 of the participant countries is not parallel. Britain, Canada, France, Italy and the Netherlands will not require the vessel until the late 1990s at the earliest, whereas the Federal Republic of Germany and Spain will need a ship much sooner. Moreover, there are differences in scale in the vessels in production. The British, Italian and Canadian vessels all weigh in at between 4000 and 4500 tonnes, whereas the Dutch, French and West German ships are all between 3000 and 3500 tonnes. As noted above, the participants in the programme all recorded a judgement that a vessel of 3500 tonnes could meet their future requirements during the 1981 NIAG feasibility study. This was reached partly in consideration of the fact that the trend in changes in equipment technology and design was reducing the size and weight of onboard systems. It does suggest an odd paradox, namely, that the part of the vessel on which common agreement is most feasible—the hull design and size—is the only part on which collaboration is unnecessary, since all of the countries involved are well capable of building their own hulls.

The question of the on-board systems for the NFR-90 in itself raises some interesting questions about the possible future shape of the procurement process. However, there are few if any of these which represent genuine co-development as it has been defined in this monograph.

Air defence systems for the NFR-90

The co-development of the NFR-90 has become closely linked in the minds of the potential customers with the development of the on-board equipment and, in particular, the air defence system. In January 1988, the British Government made the link explicit with the decision to proceed with the project definition phase only on condition that the timetables for ship production and weapon system fitting were brought into alignment at 'an early stage'.[31] There are a series of programmes under way which have been discussed in the context of the NFR-90 and its possible requirements including the NATO Local Area Missile System (LAMS), NATO Anti-Air Warfare System and the Family of Missile Systems.

The air defence system for the NFR-90 is part of a much wider discussion of NATO's integrated air defence initiated in 1979 with the 'Refined Programme for Air Defence' study to which many NATO bodies have already contributed and will continue to contribute. Discussion of the NATO air defence programme therefore extends well beyond the scope of a proposed equipment fit for the NFR-90 since it includes consideration of 'all measures designed to nullify or reduce the effectiveness of hostile air action . . . . it is directed against any type of manned and unmanned aircraft and missiles'.[32] This does not concern us here; of interest is the relationship between current air defence equipment programmes and the progress of the NFR-90.

The NAAWS programme is not directly linked to the development of the NFR-90, but has evolved from the US Navy Short Range Anti-Air Warfare System (SRAAWS) programme, a lighter and less sophisticated version of the Aegis system intended to be fitted to smaller vessels in the US Navy, in particular the FFG-7 frigate. A six-nation consortium from Canada, the Federal Republic of Germany, the Netherlands, Spain, the UK and the United States signed an MOU in Washington in October 1987 on the 'concept exploration phase' of a NATO naval air defence system.[33] Three multinational industrial teams, each led by a US company, have subsequently emerged to participate in the concept exploration phase. The teams are composed as follows: *(a)* General Electric and General Dynamics (USA), British Aerospace (UK), Marconi (UK), Signaalapparaten (Netherlands), Inisel (Spain), Siemens (FRG) and Thomson CSF (France); *(b)* Westinghouse and McDonnell Douglas (USA), Marconi and MEL (Canada), Contraves (Switzerland), Dornier (FRG), Philips (Netherlands), Bazan (Spain), Ferranti and Vickers Shipbuilding and Engineering (UK); and *(c)* Martin Marietta, ITT, Lockheed, Hughes Aircraft and United Technologies (USA), Krupp Atlas and AEG (FRG), Hollandse Signaalapparaten (Netherlands), Bazan (Spain) and Plessey (UK).[34]

The Local Area Missile System is perhaps the programme most explicitly connected to the NFR-90 programme, itself growing out of a NATO Industrial Advisory Group study.[35]

France has sought partners to share the cost of developing a Family of Air Missile Systems.[36] The family of weapon systems is in part a development of the French and Italian requirement to replace the Standard missiles in use with their Navies. In June 1988 the national armament directors of France, Italy, Spain and the UK signed an MOU covering a feasibility study into the development of a FAMS.[37] A 'family' of weapon systems is an approach developed in the United States to eliminate competitive development. In the planning phase, weapons are aggregated by type—e.g., surface-to-air or air-to-surface missiles. Then multinational consortia would be allocated responsibility for weapons of a similar type, for example, short-range weapons or medium-range weapons. There would be no co-operation in production but complete transfer of technical data.[38] In the United States the idea of a family of weapons has been restricted principally to air-to-air missile systems, but the idea has been given an application in the sphere of naval weapon systems by the French FAMS concept.

This family would consist of missiles with shared characteristics but able to be adapted for the defence of land installations as well as sea-based platforms, the feasibility study for which was initiated in 1983. The family would be based on existing national programmes, notably the French Aster 15 and Aster 30 surface-to-air missile systems, the EMPAR (European Multifunction Phased Array Radar) radar system developed in Italy and the British Sea Wolf missile.[39]

Companies from Italy, Spain and the UK have agreed to participate, and this system may become one of the competitors for the NFR-90 anti-air warfare requirement.[40] The companies involved are Thomson-CSF and Aérospatiale of France, Selenia of Italy, Ibermisil of Spain, and British Aerospace and Marconi of the UK.[41] Each of the four participant countries is contributing an equal share towards the funding of an 18-month feasibility study. The naval applications of the FAMS have been given the initial priority within France's own procurement programme, the development of the family having been divided into three phases, the first of which will be a naval point defence system.[42]

As noted above this project has grown out of existing programmes in France and Italy, but it remains a candidate as an on-board system for the NFR-90. This characteristic is shared by all of the proposals for the air defence systems for the NFR-90 described above, which relate to the combination and integration of systems already either in production or at least at the stage of prototype. The prospective programme initiated by the signature of the MOU in Washington seems to be based on the idea of developing a version of the Aegis system already installed on US Navy vessels, and the French FAMS proposal also centres on missiles, data processing and sensors in production. It is the contention here that they represent co-development because although the components are of national origin the overall package or system which is the end product is useless if any of the national components are withdrawn.

The Rolling Airframe Missile

This is the naval air defence system which perhaps comes closest to the definition of co-development. Although the origins of the programme can be traced back to embryo feasibility studies in the United States in the early 1970s, FR Germany became a major participant in the programme at its development stage in July 1976 with the signature of a Memorandum of Understanding.[43] In July 1977, General Dynamics (Pomona Division) of the United States became the primary contractors for the programme, but the West German contribution has not been cosmetic. In addition to funding 48 per cent of the programme, the same percentage as the United States (Denmark has a 2 per cent funding obligation but is currently taking no part in the programme), FR Germany will perform 55–65 per cent of the work on the control and launch system. The work will be undertaken by Ram Systems, a joint venture company including Bodenseewerk Geratetechnik, AEG Telefunken, Messerschmitt-Bölkow-Blohm and Diehl.[44] The German participation was defined in a second Memorandum of Understanding with the United States, signed in 1979. The system entered production to produce prototypes for testing in 1986 and 1987 and is planned to enter full production in 1990.

As noted above, at first glance the RAM programme represents co-development. However, there are some basic inequalities built into the structure of the programme. These inequalities are reflected in the contractual agreements between the component parts of the development programme. The West German producer can compete for only 70 per cent of the total US Navy requirement for the missile, a number which may reach 7500 missiles. The US producer can compete for all European orders. In the US, all of the data rights are owned by the Government, which can award production contracts to any company it chooses. In theory, therefore, the West German research element in the system can be given to a range of US companies which can compete against the West German production for orders. In FR Germany there is no alternative producer to the company Ram Systems, but if there were there is no question of the West German Government sharing US research and technical data with companies of its choosing.

All of these planned air defence systems are candidates for the NFR-90 equipment contract. However, the procurement of an air defence missile system and other on-board systems has not been explicitly and closely integrated with the NFR-90 programme. This lack of co-ordination has been the focus for much of the criticism aimed at the NATO frigate programme. This vagueness is illustrative of the complexity involved in co-ordinating the design and development of multinational systems of the size and scope of a major naval vessel. In the 1988–89 edition of *Jane's Fighting Ships,* the editor notes:

. . . the chances of several navies all acquiring an air defence ship at much the same time are not high even if agreement can be reached on . . . competing area defence surface to air missiles which are still being developed. It is difficult enough to get sensible agreement within a single Ministry on a weapons system of that complexity; add to that a factor of eight and the NFR-90 project could be said to be overambitious.[45]

The governments involved in the NFR-90 project themselves have recognized that the discussion of the co-operative development of weapon systems may not reduce the extent of the arms trade between members of the major alliances. In its discussion of the 1988 Statement on Defence Estimates, the United Kingdom Defence Committee noted that 10 per cent of the Ministry of Defence's procurement expenditure is now spent abroad, an increase from 5 per cent in 1985, whereas the volume of procurement accounted for by collaborative work has remained stable at around 14–16 per cent. A Deputy Under Secretary responsible for procurement noted: ' . . . if we are going to have competition we have increasingly to look at the cheapest and most effective ways of buying equipment and meeting our requirements and that must mean, if our costs and timescales for requirements are met, that we can look overseas as well as in this country'. Commenting on his evidence, the Committee reported: 'Continuing pressure on the defence budget may nevertheless be a powerful stimulus to overseas purchase . . . this greater willingness to consider overseas purchase may signal a change in policy towards the defence industrial base'.[46]

Consequently it may be that co-development and co-production are not yet alternatives to the more traditional forms of arms transfer, direct transfers of finished systems or sub-systems and the sale of licences.

Notes and references

[1]   Chisholm, J., 'Opportunity and challenge', *Armed Forces*, July 1988, p. 327.
[2]   Sabin, P. A. G., 'World War 3: A historical mirage', *Futures,* Aug. 1983, p. 273. The extent to which nuclear disarmament will reverse these perceptions is the subject of a SIPRI research project Security Without Nuclear Weapons? conducted by Regina Cowen Karp.
[3]   The amendment is reproduced in *NATO's 16 Nations*, July 1985, p. 23.
[4]   Roth, W. V., 'After the Nunn–Roth Amendment', *NATO's 16 Nations*, July 1985, pp. 20–21.
[5]   *Defense News*, 8 Feb. 1988, p. 6.
[6]   Lenorovitz, J. M., 'France planning more joint programs', *Aviation Week & Space Technology*, 30 May 1983, pp. 83–84.
[7]   Houwelingen, J. V., 'Towards European defence industry', *NATO's 16 Nations*, Oct. 1985, p. 45.
[8]   Houwelingen (note 7).
[9]   *Jane's Fighting Ships 1988–89* (Jane's: Coulsden, 1988).

10 Lenton, T., 'French naval construction', *Maritime Defence*, Oct. 1982, pp. 375–77; de Blocq van Kuffeler, F., 'Update on construction: the Netherlands Navy', *Navy International*, Jan. 1985, pp. 16–22.

11 de Blocq van Kuffeler, F., 'Royal Netherlands Navy Defence Plan', *Navy International*, Apr. 1984, pp. 198–202.

12 'New Dutch MCMV design for export', *International Defense Review*, Dec. 1984, p. 1792.

13 'The future size and role of the Royal Navy's surface fleet', House of Commons *Defence Committee*, Sixth Report, Session 1987–88, HMSO, 21 June 1988, p. xxii; Cook, N., 'US/UK torpedo agreement soon', *Jane's Defence Weekly*, 4 June 1988, p. 1087.

14 *Jane's Defence Weekly*, 28 Jan. 1989, p. 148.

15 *World Missile Forecast* (Forecast Associates Inc: Newtown, Connecticut, 1988); *Jane's Weapons Systems 1988–89* (Jane's: Coulsden, 1988), p. 467.

16 Modular design means that weapons and electronic systems are fitted into pallets with standard dimensions, supports and connections to power supplies and data cables. These pallets can then be lifted into the ship hull and connected once installed on board. All Western European and US weapons can be fitted into one of four standard sized pallets.

17 Smith, L., 'Is the NATO frigate for real?', *Armed Forces Journal*, Nov. 1984, pp. 109–10.

18 '1986 crucial year for NATO frigate programme', *International Defense Review*, May 1986, p. 552.

19 *Atlantic News*, no. 1763 (6 Nov. 1985).

20 The United States could not insist that the successful company (Westinghouse Electric Corp.) join the feasibility study. NATO has its own procurement procedures, Periodic Armaments Planning System (PAPS), which is different in many respects from US national procedure. For a discussion of the differences see Kelley, W. E., 'A NATO frigate for the 1990s', *Proceedings of the US Naval Institute*, Mar. 1985, p. 184.

21 'Statement of Intent for NFR-90', *Naval Forces*, July 1986, p. 139.

22 'NFR-90—White Elephant or White Hope?', *Defence*, Dec. 1987, p. 739.

23 'The future size and role of the Royal Navy's surface fleet'(note 13).

24 Bloom, B., 'US rejoins NATO warship study', *Financial Times*, 1 May 1984.

25 Truver, S. C., 'Tomorrow's Fleet', *Proceedings of the US Naval Institute*, Naval Review issue 1988, p. 313.

26 City class refers to ships of phases I and II of the SRP. The phase I ships are of the Halifax class, phase II ships will be called the Montreal class. For more information on Canada's air defence escort see Shadwick, M., 'The Canadian Navy: recovering from rust out', *Naval Forces*, Feb. 1988, pp. 66–67. For the political and economic backdrop of the extended procurement decision see Middlemiss, D., 'Canada and defence industrial preparedness: a return to basics?', *International Journal*, autumn 1987.

27 'Canada to build new frigates', *Financial Times*, 5 Jan. 1978.

28 *Jane's Fighting Ships 1988–89* (Jane's: Coulsden, 1988) p. 173; Preston, A., 'France's new air defence ships', *Defence*, June 1988, p. 443.

29 'France announces plan for 12 new frigates', *Jane's Defence Weekly*, 30 Apr. 1988, p. 821; *Defense News*, 18 Apr. 1988, p. 34.

30 *Jane's Fighting Ships 1988–89* (Jane's: Coulsden, 1988), p. 207.

31 'The future size and role of the Royal Navy's surface fleet' (note 13), p. xxii.

82   *The Naval Arms Trade*

32 Altenburg, W., 'Defence in the air, NATO's integrated air defence today and in the future', *NATO's 16 Nations*, Aug. 1986, p. 22. Altenburg explains the need to consider active air defence as including an integrated response by fighter aircraft, anti-aircraft guns, surface-to-air missiles and electronic counter-measures.

33 Fairhall, D., 'Whitehall limits NATO air defence agreement', *The Guardian*, 23 Oct. 1987; 'NATO anti-air consortium', *Defence*, Dec. 1987.

34 *International Defense Review*, Jan. 1988, p. 16.

35 *International Defense Review*, Mar. 1988, p. 226–28.

36 *Jane's Weapons Systems 1987–88* (Jane's: London, 1987), pp. 490–91; Grove, E., 'Western Europe's Navies', *Naval Forces*, Feb 1988, p. 74.

37 'Collaborative naval SAM proposed', *Defence*, Mar. 1988, p. 380.

38 A more detailed overview of the concept is contained in a report prepared for the Under Secretary of Defense for Research and Engineering in December 1980. Kanter, H. and Fry, J., *Cooperation in Development and Production of NATO Weapons: An Evaluation of Tactical Missile Programs*, Institute for Defense Analyses, IDA report R-253.

39 Marsh, D., 'Anglo-French tie proposed for anti-missile system', *Financial Times*, 11 July 1986; *Aviation Week & Space Technology*, 8 Dec. 1986, p. 13; *Jane's Weapons Systems 1987–88* (Jane's: London, 1987), pp. 490–91.

40 Friedman, N., 'World naval developments 1987', *Proceedings of the US Naval Institute*, May 1988, p. 224.

41 Hewish, M., 'Trends in shipborne radar', *International Defense Review*, June 1988, p. 669.

42 *Jane's Defence Weekly*, 15 Apr. 1988, p. 919; and *Defense News*, 25 Apr. 1988, p. 4.

43 *Jane's Weapons Systems 1988–89* (Jane's: Coulsden, 1988), p. 480.

44 *World Missile Forecast* (note 15).

45 *Jane's Fighting Ships 1988–89* (Jane's: Coulsden, 1988) p. 109.

46 Seventh report of the defence committee, session 1987–88, *Statement on the Defence Estimates 1988*, 28 June 1988, p. xix.

5. Australia and New Zealand

I. Security perceptions in Australia and New Zealand

Australia and New Zealand share both a geographical isolation on the fringes of Asia and the Pacific and the fact that large sections of their populations originated in Europe. In both cases they exchanged a collective security system based on the British Empire and later the Commonwealth for an alliance with the United States as the central feature of defence policy. Forces from Australia and New Zealand participated in both World Wars and subsequently in wars on the Asian mainland, both in Korea and in particular in Viet Nam. Both Australia and New Zealand have revised their security relationship with the United States in the light of a judgement that the ability and willingness of the US to engage in a war on the Asian mainland has diminished. They have come to emphasize increased self-reliance and greater co-operation with one another.

Australia and New Zealand have also revised their attitudes to regional security. Traditionally, both have defined a significant role for their armed forces on the Asian mainland. The military manifestation of this role has been their commitment to the Five Power Defence Arrangement (FPDA), along with Malaysia, Singapore, and the UK, which became operative in 1971. The Australian commitments under the FPDA have included the permanent stationing of a significant proportion of Australia's tactical air forces at the Butterworth Air Force Base in Malaysia. Australian forces have gradually been drawn back from Malaysia and Singapore since 1973, and in 1987 it was announced that the permanent presence at Butterworth would give way to a rotational deployment of Australian F/A-18 aircraft.[1] Australian commitments to South-East Asia will continue to include the stationing of long-range surveillance aircraft at Butterworth, and Australia will continue to play a major role in the upkeep and maintenance of the base. Australia withdrew its forces from Singapore in 1973; New Zealand maintained a 700-man force there until the announcement in the 1987 Defence White Paper that this force would be drawn back to New Zealand to form a Ready Reaction Force (RRF) relevant to the needs of the region and ready to respond to military contingencies.[2]

Although these factors have had an important impact on foreign and security policy in both countries, there are also some significant differences between Australia and New Zealand. Australia has traditionally included more of a northern orientation in its foreign and defence policy while New Zealand has focused on the south-west Pacific. In spite of its small population, much of which is concentrated in the south-east and south-west of the continent, Australia has seen itself as an important regional power in South-East Asia. New Zealand, with significant indigenous Maori and island populations, has had a slightly different orientation.

In spite of their differences, it is appropriate to deal with Australia and New Zealand in one chapter because of the increasing co-operation between them in the area of economic and industrial development. This co-operation has at its root the reorientation of trade from Australia and New Zealand during the 1970s and 1980s. Increasingly both countries are focusing on expanding trade with the countries of east and South-East Asia and with the United States. Australia and New Zealand have had to make a gradual and structural adjustment to the formation and evolution of the European Community (EC) and in particular to increased British involvement in the EC.[3] Australia and New Zealand have moved towards establishing a single market in goods and services by 1990 through the Closer Economic Relations (CER) agreement.

In Australia and New Zealand foreign and defence policy has become meshed with changes in attitudes to alliance membership and perceptions of regional security policy. There already exist a large number of interlocking sub-regional organizations and institutions in East and South-East Asia. Meanwhile Japan has tried to promote the concept of the Pacific Rim to focus on finding an institutional framework that can bring together diverse cultural and political groups basically for the purposes of expanding trade.[4]

Co-operative naval armaments programmes are one area in which Australia and New Zealand have been able to formulate joint responses to these regional developments. However, in Australia in particular, the review of defence policy has led to the aspiration significantly to expand the defence industrial base.

II. Australia

Equipment programmes in Australia

All of the major current naval or associated procurement programmes in Australia include a substantial percentage of local construction. These programmes include the purchase and construction under licence of six Type 471 submarines from Sweden, the construction under licence of two FFG-7 Oliver Hazard Perry class frigates of US design and the licensed construction of eight light patrol frigates for the Royal Australian Navy (RAN) and an as yet to be determined number for New Zealand. There is a programme to refit Oberon class submarines and P-3C surveillance aircraft with versions of the US RGM-84A Harpoon missile. There has also been a major focus on the maritime air component of the Australian armed forces. The centre-piece of this has been the purchase and subsequent licensed production by the Royal Australian Air Force (RAAF) of 75 F/A-18 Hornet aircraft in various versions. The purchase and licensed construction of US SH60 Sea Hawk helicopters for deployment on both FFG-7 class frigates and on the light patrol frigate to be built for both the Australian and New Zealand navies has been planned, together with the conversion of the existing fleet of Boeing 707 aircraft to tankers for use in in-flight refuelling, modifications being carried out by Israeli Aircraft Industries. There has already been a major investment in an intelligence and surveillance network based on the expansion of the Jindalee 'over-the-horizon' radar network in northern Australia.

The chronology of major decisions relevant to these Australian programmes stretches back to a 1972 Australian Defence Review and a 1976 Defence White Paper, both of which emphasized the need to develop a more self-reliant approach to defence. Plans to build in Australia a successor to the destroyers in service with the RAN in Australia had been abandoned in 1974 after a large expenditure on assembling a design and production team. Instead, the Department of Defence (DoD) decided to buy outright four FFG-7 Oliver Hazard Perry class frigates from the United States.[5] However, in 1983 a contract with Todd Pacific Shipyards of San Pedro, California was announced to license the Australian government-owned Williamstown Naval Dockyard (Willdock) to construct two FFG-7 class frigates for Australia.[6] Licensed production of these vessels was not contractually linked with the earlier purchase of frigates, and the decision to build decision to build ships of this tonnage in Australia surprised analysts since the dockyard had not completed a major ship since 1971. Moreover, the construction of the FFG-7 has been extremely problematic because of a succession of industrial disputes.[7] The dockyard has since been bought by the Australian Marine Engineering Corporation (AMEC), which is one of the prime contractors short-listed to build the Australian light patrol frigate. The decision to produce the FFG-7 fits with the tone of the 1976 Defence White Paper, which stressed the need for greater self-reliance in the wake of the Viet Nam War and the reassessment of US foreign policy implied by the Guam Doctrine. These twin developments threatened the traditional Australian asumption that security was indivisible from that of its superpower ally. Not only was it uncertain in the mid-1970s that the United States would maintain the regional forces necessary to respond to security threats directly, but it was possible that the Guam Doctrine would involve the transfer of military equipment to South-East Asian countries with which Australia might, in certain circumstances, be in confrontation. Secretary to the Department of Defence W. B. Pritchett spoke of the Australian Government's dilemma in reconciling 'competing demands between our national defence posture and the international policy for the support of our ally'.[8]

The procurement programme currently under way in Australia has its roots in decisions taken in the 1970s. However, there have been significant changes in Australian policy during the 1980s. As Andrew Mack puts it:

In Australia the need for greater defence self-reliance was hinted at in the 1972 Defence White Paper, signalled clearly in the 1976 White Paper, but not articulated with any real degree of strategic coherence until the publication of ministerial consultant Paul Dibb's 1986, Review of Australia's Defence Capabilities, and the Hawke Labor Government's subsequent 1987, Defence White Paper.[9]

There have been three significant shifts in Australian thinking on defence. First, there was an increasing northern and north-western orientation in planning and in particular an increased focus on the South-East Asian archipelagos and the Indian Ocean. Second, there was a tendency to focus more on domestic dimensions of defence policy. This has meant in particular the expansion of the defence industrial base in an effort to increase Australian defence capabilities while enhancing the domestic technology base. Third, there was the encouragement by Australia of the development of a regional security system. This does not imply an Australian role as a 'regional policeman', but is more based on co-operative programmes in the spheres of procurement, reciprocal ship and aircraft visits (including the operational use of airfields), joint training programmes and political and diplomatic support on issues of joint interest.

The Dibb Report

In Australia, a review of defence policy found expression in a major Government document, published in 1987.[10] The Defence White Paper was preceded by a major defence review by the Labour Government covering all aspects of Australian defence which culminated in the preparation of two reports. The first has become known as the Dibb Report after its author, Paul Dibb, and looked at the strategic environment of Australia. The second looked at the existing defence industrial base in Australia.[11] The Dibb Report questioned the relevance of traditional military commitments—such as joint operations in support of ANZUS allies and the protection of sea lanes in a major conflict—to the specific military environment in South-East Asia, although the first of these commitments had been effectively abandoned in 1976 in a previous DoD policy paper.[12]

The growing northern orientation of Australian policy has led to a major defence reorganization which includes the construction of new facilities to allow the relocation of sizeable portions of Australia's military assets to *HMAS Stirling*, near Perth, and the construction of a new air base on the Cape North Peninsula, due to become operational in the 1990s.[13] A debate has also reopened on the question of the future value of the Christmas and Cocos Islands, Australia's north Indian Ocean territories located close to the Indonesian island of Java, given their potential value in extending the range of operation of Australian P-3C and F/A-18 aircraft. As recently as 1986 a former Chief of the Air Staff stated:

It would be somewhat naive to assume that the status of being Australian territory automatically makes these far flung possessions a practical proposition for a smallish military power like Australia. Such catchcries as 'we will not surrender an inch of Australian territory' might be tolerable political rhetoric when no threat darkens the horizon. They could be totally irresponsible when the defence of such places was clearly impractical . . .[14]

The Dibb Report sparked a great controversy within the Australian armed forces first because it was alleged to be 'isolationist' in tone, and second because of the implications of a section which remains classified, apparently to avoid any deterioration in relations with Indonesia. This was a discussion of contingencies which might lead to an Australian intervention in Papua New Guinea. The subsequent Defence White Paper presented to Parliament in March 1987 was different in tone, presenting changes in Australian policy as evolutionary and stressing the maintenance of traditional ties with allies.

In response, the armed forces stressed the need to retain traditional elements of Australia's forward defence strategy based on engaging potential threats a long way from the mainland rather than allowing them to approach before responding. The review touched areas of inter-service rivalry in that the Navy, in particular, wanted to retain capabilities to operate a long way from Australia's territorial waters, rather than seeing the primary task of engaging surface vessels pass to the Air Force.[15] However, the review actually fitted into, rather than departed from, the trends under way in Australian force planning.

A symbol of change was the decision in 1983 by the Labour Government not to replace the aircraft-carrier *HMAS Melbourne*. Given that defence of the carrier and operating with a carrier group were the primary roles for the surface forces in the Australian Navy, the decision not to replace the *Melbourne* was a clear signal of shifting priorities in force planning. In terms of funding for maritime aircraft, priority has been given to the acquisition of the F/A-18 fighter and the refurbishment of the existing force of F-111 bombers, the procurement of long-range maritime patrol aircraft armed with long-range anti-ship cruise missiles and the creation of a tanker fleet equipped for in-flight refuelling. Attention has also been paid to the creation of a force of airborne surveillance platforms. The airborne early-warning systems would be under the operational control of the RAAF and seem likely to consist of a mix of Australian aircraft already in service, refitted with imported radars and electronic systems, on the one hand, and new platforms bought as dedicated early-warning aircraft, on the other.[16] These aircraft would supplement the creation of a comprehensive network of land-based surveillance radars based around three Jindalee type over-the-horizon radars. These radars are claimed to have the capability to monitor air and shipping movements up to 3000 kilometres from Australia's northern coastline.[17]

Within these plans, the recent naval procurement decisions taken by the Australian Government offer a great deal of evidence of the hierarchy of priorities within Australian defence policy. The Defence of Australia identifies three primary tasks relating to Australian security: first, preventing an adversary from making use of maritime approaches to Australia; second, surveillance of the enormous area of sea and airspace relevant to Australian security; and third, the ability to project force within an 'area of direct military interest' stretching 1000 nautical miles beyond the coastline.[18] Defence Minister Beazley suggested that the strategy called for an Australian defence force capable of meeting 'any hostile force within our area of direct military interest with successive layers of forces capable of detecting, identifying and engaging any hostile approach'.[19] In terms of budget priority, the modernization of the submarine fleet through the retro-fitting of existing Oberon class submarines and the subsequent purchase of new submarines of Swedish design appears to have been given greater priority than major production of surface vessels.

The projection of power within the region has been accorded lower priority within the strategy laid out by the Australian Government, and the procurement programmes under way or planned for the next decade are principally designed to meet the first two priorities. The establishment of priorities among arms import progammes is a consequence of the fact that the announcement of new equipment programmes occurred almost simultaneously with an announced reduction in military expenditure.[20] This apparent paradox was explained by Treasurer Paul Keating as a reflection of the long-term nature of all Australian procurement programmes. The White Paper was prepared on the assumption of no long-term growth in funding and possible cuts. All of the programmes laid out in the White Paper were expected to unfold over a minimum of 15 years.[21]

There is no evidence in the development of Australian programmes that the Government assigns high priority to force projection within the 1000-nautical-mile zone defined in the White Paper except in collaboration with regional countries against extra-regional intrusions.[22] There are no published plans for additional amphibious capabilities or major increases in airlift capacity. At the time of the coup in Fiji, Australian policy was restricted to economic and political sanctions rather than any attempt at intervention, although there was a considerable Australian presence in the vicinity of the island at the time of the first coup in May 1987. The frigates *HMAS Sydney* and *HMAS Adelaide* arrrived in Fiji on the day of the coup, while the patrol boats

Wollongong and *Cessnock* arrived the following day. The destroyer *HMAS Stalwart* was a little further north and *HMAS Parramatta* was also relatively close and was ordered to approach Fiji's territorial waters.[23] Moreover, the Australian and New Zealand governments received requests for assistance from the Prime Minister of Fiji, recognized by both countries as the legitimate head of the Fiji Government.

The Australian Type 471 submarine programme

The highest-profile single naval arms deal involving Australia has been the decision to order a new generation of conventional attack submarines from Sweden. The Australian Navy currently operates Oberon class submarines bought from the UK in the 1960s and 1970s, which are currently being modernized, a programme that includes the fitting of a version of the Harpoon anti-ship missile purchased from the United States. However, these vessels will begin to be retired during the 1990s, and a new generation of Australian submarines was selected in 1987.[24]

The replacements will be six Type 471 submarines to be designed and built by a consortium headed by Kockums AB of Sweden and Rockwell Collins of the United States. All of the boats are to be assembled in Australia rather than beginning with delivery of complete boats from Sweden.[25] The timetable for the replacement programme as announced by the Treasurer to the Australian Parliament envisaged beginning construction on the first submarine in 1989 with a view to launching in 1995.[26] The contract is extremely complex and involves an enormous number of sub-contractors which may ultimately reach over 100 companies. The Australian Submarine Corporation will have overall responsibility for the programme and for building the boats themselves. The corporation consists of Kockums, Wormald International Australia (in spite of its name, a US company), and CBI Constructors (also of the United States). The corporation also includes a government statutory body, the Australian Industry Development Corporation, which will represent the interests of Australian companies.[27] All of the combat systems on the submarine will be provided by a consortium including Rockwell International and Singer Librascope Division of the United States, Thomson CSF of France, Ship Systems Australia, Computer Sciences of Australia and Scientific Management Associates. Ship Systems Australia is itself a consortium including Fairey Australasia, the British companies British Aerospace and Plessey, and Rockwell Electronics of the United States. In order to co-ordinate such a complex company, the Australian Submarine Corporation employed Bath Iron Works of the United States as consultants, Kockums having no experience of exporting submarines or of managing a programme of this complexity. As noted above, the Australian Government decided to build all the submarines in Australia, although in the light of past shipbuilding experience the contract with the Australian Submarine Corporation includes a caveat that production in foreign yards will be considered for subsequent ships in the programme if construction costs exceed the parameters laid out in the contract.[28]

The structure of the submarine contract is indicative of the increased emphasis in Australia on expanding the defence industrial base. Subsequent to the Cooksey Report, the Department of Defence published a new Defence Policy for Australian Industry which laid out a need to expand the defence industrial base both as a complement to increased self-sufficiency and to use defence resources for the development of local industry. As a second stage, Australian industry would, in combination with the Government, seek overseas markets for Australian defence goods.[29]

Australian–New Zealand collaborative naval procurement

Along with the submarine programme, another major naval procurement decision which will contribute to the development of the Australian defence industrial base is the planned construction of a new frigate by Australia in collaboration with New Zealand.

The earliest delivery of the planned light patrol frigates is expected to be the 1990s. In January 1987, Australian Defence Minister Beazley invited shipbuilders to submit tenders on the basis that the ships would have to be constructed in Australia. The programme had a clear industrial and economic imperative alongside the military requirement. Beazley was moved to comment that 'the Navy is offering the salvation of the Australian shipbuilding industry'.[30] New Zealand joined the programme in March 1987 with the signature of a Memorandum of Understanding by the respective Defence Ministers. Given that co-ordinated procurement is impossible between countries with totally antagonistic relationships, joint programmes are likely to fulfil political and economic functions in addition to adding to military capabilities. Joint procurement of a new generation of escort vessels by Australia and New Zealand has been indirectly influenced by the 1983 Closer Economic Relations (CER) Agreement, intended to produce complete freedom of movement of all goods between the two countries by 1995, which has been interpreted as a move towards integration of the Australian and New Zealand economies.[31] The agreement, not dissimilar to the Single European Act within the EC, should allow completely open competition between Australian and New Zealand companies for public contracts. Once Australia and New Zealand agreed on a joint purchase of the same system, the opportunities for New Zealand companies to participate in the construction was enhanced.[32]

Two consortia, AMEC and AWS (Australian Warship Systems), were awarded design and development contracts for the construction of the patrol frigate. The bid from AMEC was based on licensed production of the West German MEKO-200 and that from AWS on the Dutch M-class frigate.[33] As noted above, the AMEC consortium includes the company that has bought the Williamstown dockyard which is building FFG-7 frigates for the Australian Navy. The terms of the dockyard sale included the provision that, should AMEC win the contract, the vessels under construction must be completed before any work on the light patrol frigate begin.[34]

The patrol frigate programme has highlighted the extent of the impact that major naval contracts can have on the defence industrial base. AMEC, which began as a consortium of three companies, has been the centre of intensive take-over activity as new companies have joined the group and predatory outsiders have tried to buy members of the consortium to secure a share of the final contract should AMEC win it. The AMEC group now includes interests from the USA, Malaysia, Indonesia and Israel as well as Australia and New Zealand.

The frigate programme also became entangled in domestic Australian politics, and in particular in the relationship between the central government and respective state governments of New South Wales and Victoria. The AWS group is based in New South Wales, while AMEC is located in Victoria. At times during the competition there was some speculation that the potential electoral consequences of awarding the entire contract to one or other state may lead to production being split between shipyards. A further complication in managing the overall contract has been the uncertainty over the full extent of New Zealand participation.

Although the Australian Government defined its requirement for new patrol frigates independent of any New Zealand programme, the two governments took advantage of the fact that they both needed a similar ship at the same time. During the debate in New Zealand on the 1983 Defence Review, it was proposed that frigates due for modernization should be replaced not with new, equivalent systems but with submarines. The 1987 Defence White Paper explicitly rules out the purchase of submarines, stating that a submarine force 'would have limited value in meeting the more realistic circumstances that call for the use of naval force'.[35] In the event, the 1983 Defence White Paper also expressed the view that two Leander class frigates, *Waikato* and *Southland* (both in service since the mid-1960s), should be replaced with new rather than second-hand frigates, but that these new vessels would be smaller than the ships they were replacing. In the past, modernization of the Royal New Zealand Navy has meant turning to Europe or the United States. However, this would almost certainly require the purchase of a ship that went beyond New Zealand's perceived requirements. New Zealand has outlined a requirement for a ship able to stay at sea for long periods (to patrol an enormous sea area), with surveillance and communications capabilities as well as basic air defence systems, a limited capability against submarines and an ability to accommodate a light helicopter. These characteristics could be met by a light frigate without expenditure on the scale required to buy a vessel such as that on offer from traditional European suppliers.[36] Consequently, New Zealand investigated joint acquisition of such a ship with Australia, which has similar mission requirements.

In 1988 New Zealand agreed in principle to purchase four frigates and, in addition to the eight required by Australia, the series production would have been twelve. However, in 1989 there were suggestions that the Royal Australian Navy may have a requirement for twelve vessels itself. Moreover, a considerable number of members of the New Zealand Government supported New Zealand's withdrawal from the programme altogether or at least a scaling down of the size of its commitment.[37] One alternative proposal was the purchase of a Danish design, the Stanflex, basically an advanced and spacious offshore patrol vessel (OPV) which could later be refitted with a variety of armaments should the New Zealand Government so decide. In August 1989 the Australian Government decided to purchase frigates based on the West German MEKO-200 design to be built in Melbourne.

III. New Zealand

In New Zealand, current programmes involve modernizing P-3C Orion reconnaissance aircraft with new surface search radars; extending the range of existing A4 Skyhawks by creating a fleet of in-flight refuelling tankers and extending the life of these aircraft with a modernization programme; the acquisition of light patrol frigates in conjunction with Australia (discussed above); and the acquisition of a logistics support ship for New Zealand's Ready Reaction Force.

In 1987 the New Zealand Department of Defence produced a Review of Defence Policy.[38] Prime Minister Lange called the document 'the most fundamental change in defence policy since World War II',[39] contributing to the widespread impression that New Zealand's attitudes to defence were revolutionary rather than evolutionary. Certainly, the philosophy of New Zealand's security policy since the Labour Party took power in 1984 has been markedly different from that of Australia in spite of their geographical proximity (a relative term in the south-west Pacific) and common membership of the ANZUS alliance. In August 1985 the United States stopped the transfer of technical information to New Zealand after Wellington required US Navy ships to confirm or deny the presence of nuclear weapons or propulsion systems on board. In other programmes, New Zealand policy has been less severely affected. In 1986, the US suspended its observation of alliance obligations as far as New Zealand is concerned. Procurement policy has certainly not been immune to these changes. However, procurement policy illustrates both the pre-1984 roots of some recent changes in policy and also a degree of continuity in some programmes in spite of the changes in US–New Zealand relationship.

New Zealand was forced into a reappraisal of its defence commitments at the end of the 1970s as a result of the combined economic impact of the loss of traditional markets and the oil price rises, and in 1978, with no direct threats to New Zealand being defined, politicians of all parties considered whether radical restructuring of the fleet could meet New Zealand's needs at reduced cost. Resource-related themes dominated New Zealand thinking on defence in the late 1970s, notably the ability to devote resources to defence in the face of economic difficulty and the need to increase the domestic industrial input in defence procurement. Certainly after 1980, when narrow defeat at the general election marked the arrival of Prime Minister David Lange and the Labour Party as a genuine alternative to the National Party, domestic politics were a central component of the defence debate alongside technical and military factors and alliance considerations. Therefore, the need for self-reliance in defence stressed in the 1987 defence review is a theme that had its roots in the 1970s.[40]

New Zealand and regional security

In terms of their impact on procurement policy, the most important movements in New Zealand defence policy have been the growing divergence of views on regional issues between the United States and New Zealand, on the one hand, and the growing political and economic co-operation between Australia and New Zealand, on the other. The theme of self-reliance in part reflects the fact that the defence review has taken place alongside changes in the nature of the alliances in which New Zealand participated and changes in the regional political environment. Prior to 1984, New Zealand security policy was in many ways a function of alliance membership. In 1985–86 the full scope of reliance on its allies, and in particular the United States, for operational support was laid out by the New Zealand Chiefs of Staff in a report prepared for then Defence Minister F. O'Flynn. They noted that New Zealand lacked air defence for airfields and ground forces, heavy artillery, satellite communications, over-the-horizon radar, interceptor aircraft, air-to-surface munitions, main battle tanks, and submarines. Defence intelligence was reliant on the US and, to a lesser extent, Australia.[41]

The New Zealand Government did not dispute the Chiefs of Staff military assessment; rather they rejected the premise that ANZUS was the most appropriate context in which to cast security policy. In New Zealand, naval procurement has been seen as an important part of a wider re-evaluation of attitudes to regional security and the nature of relations with the United States. For New Zealand, whose defence policy orientation has traditionally been north-east (towards the south-west Pacific) rather than north (towards Australia), the Pacific has historically been a benign environment and the question most often raised in the context of defence policy was 'defence against whom?'.

Prime Minister Lange has put this argument in a most unambiguous form, noting:

In terms of threat to our independent existence, we are in exactly the same positions in 1987 as we were in 1984. There is no threat and none can be conjured. Those who refuse to take comfort in the Government's assurances on this point can perhaps take comfort in the assertions of its most ardent critics: not even the fiercest of them has been able to identify an enemy massing on the horizon. Instead they tend to see the Government's dereliction as the little tear we have made in the seamless fabric of nuclear deterrence. In the wider meaning of security we are certainly not worse off. Those who talk of the withdrawal of the American security guarantee as if they are talking about a fact are doing, in fact, a disservice. The existence of the so-called guarantee was at the very least a matter of dispute.[42]

New Zealand programmes have already been affected by the effective dissolution of New Zealand's alliance relationship with the United States. The initial Request for Proposals (RFP) related to the contract to produce the patrol frigates described above was issued from Canberra before it was certain that New Zealand would buy the same ship. However, in framing its request, the Australian Department of Defence required respondents to indicate clearly in their proposal equipment that could not be made available to New Zealand.[43]

The modernization of New Zealand A-4 Skyhawks was initiated as a result of the 1983 Defence Review, and was under way when the break in the ANZUS relationship occurred. The management of the programme was the responsibility of a division of Lockheed, a US-based corporation. The programme included a refurbishment of airframes, in particular the replacement of the wings, which has been conducted in New Zealand. However, it also included the addition of avionics to increase the effectiveness of the aircraft in maritime strike missions and ultimately the addition of an anti-ship version of the AGM-65 Maverick missile. Most of the requisite sub-systems were to come from the US companies Emerson and Litton. After August 1985, it seemed that the programme would be suspended when the US Government withdrew its licence for technology export. However, not only has the programme been continued but New Zealand appears to have purchased a small number of Maverick missiles from Jordan in 1986, a transaction which should have received US sanction.[44]

The disruptive impact on procurement was acknowledged in the 1987 White Paper, in particular the consequences of the lapse of a

1982 Memorandum of Understanding that had been used as the basis for US Foreign Military Sales credit. (One of the programmes benefiting from US credit was the A-4 modernization programme.) Moreover, future New Zealand imports of US arms were made more difficult as a result of the passage of a bill in the US Congress in October 1987 by which New Zealand lost the preferential treatment accorded to allies under the Foreign Assistance Act and the Arms Export Control Act.[45] These actions help to clarify the US position on arms sales to New Zealand. Sales have not been made impossible, but New Zealand will in future have to pay in cash and at the full market price for equipment the transfer of which was previously eased by special financial arrangements.

Changes in US–New Zealand relations have reduced the economic incentives to buy US equipment and weakened the advantages of compatibility and interoperability. However, changes in New Zealand programmes are also evidence of an increasingly close relationship between Australia and New Zealand in the area of security policy generally. New Zealand Defence Minister Robert Tizzard expressed the new policy as follows: 'Since we have no defined enemy, we need vessels that can perform the functions of the various roles we see for ourselves. These include maintaining a presence in the South Pacific and building co-operation with Australia'.[46] Within defence establishments of the region, there has been some concern that the challenge to the traditional architecture of defence policy caused by changes in US–New Zealand relations may have come at exactly the time when regional stability was under threat. The 1987 White Papers of both Australia and New Zealand have included a discussion of the possible need for forces capable of intervention in the region. However, within their procurement programmes there are clear signs of the different attitudes of the two countries to the concept of intervention.

Closer economic co-operation between Australia and New Zealand was referred to above, but attitudes to regional security have also been characterized by a greater need for co-ordination. Australian and New Zealand attitudes to regional security have been conditioned by a perceived need to support the policies of local micro-states in cases where they are endangered by extra-regional parties. One manifestation of this has been Australian and New Zealand support for and lobbying on behalf of the Treaty of Rarotonga. Another has been the support for the South Pacific Forum Fisheries Agency (FFA). Among the formal requirements of the FFA are the promotion of regional co-ordination and co-operation with regard to relations with distant-water fishing countries and matters of surveillance and enforcement in respect of the rights and obligations accorded under the LOS Convention.[47]

In February 1987, Australian Minister for Defence Kim Beazley laid out the rationale for this policy. He noted that

a self-reliant defence posture demands that we shape our defence capabilities to suit our environment. Likewise it requires that we pay great attention to maintaining and strengthening the congenial features of that strategic environment. In doing this we must go well beyond the specific area of defence. Our policies must encompass aid, trade, immigration and a host of other issues. But as the island groups of the South Pacific have developed their own strategic perceptions and concerns, this has increased the scope for co-operation in this field.[48]

In the same statement Beazley laid out some specific regionally oriented security initiatives. These included moves to help island countries upgrade their national maritime surveillance systems, deployment of RAAF long-range maritime patrol aircraft to the region, increasing numbers of RAN ship deployments and technical support to island defence and security forces. These initiatives have included a defence agreement signed by Australia and Papua New Guinea in December 1987 and the growing number of Australian–New Zealand exercises.[49]

The Pacific Patrol Boat programme was launched in 1983 to provide a total of twelve 31.5m patrol boats to Papua New Guinea (four), Fiji (four), Solomon Islands, Cook Islands, Western Samoa, Vanuatu (one each). Tuvalu is presently considering participation in the project. Paid for by Australia, the programme also includes training and advice and has been complemented by Australian naval visits and bilateral efforts to improve communications facilities, hydrographic skills and the accurate delineation of 200-nautical-mile EEZs. Australian P-3C Orion maritime patrol aircraft have been deployed to air bases in the south-west Pacific on a regular basis since 1983. Australia has also agreed to pay for the upgrading of airfields in Vanuatu. Official statements by the Australian Government envisage increasing the number of such deployments and the participation of New Zealand patrol aircraft in a co-ordinated deployment schedule.[50] These complementary programmes were formally linked in 1984 into the Regional Fisheries Surveillance and Enforcement Project.[51]

The programme has been disrupted by the second coup staged by Colonel Rabuka in Fiji on 25 September 1987, following which Australia suspended all deliveries to Fiji, recalled defence co-operation advisers and suspended all aid with the exception of grants paid to Fiji students already in Australia.[52] However, the rift between Fiji, on the one hand, and Australia and New Zealand, on the other, may prove temporary given the efforts to strengthen existing regional security linkages.

This greater co-ordination within the region need not mean that regional military intervention is seen as a profitable or necessary exercise. In fact, the reactions to the coups in Fiji suggest that it is an extremely unlikely option for either the Australian or the New Zealand governments. However, such intervention has not been ruled out by either Australia or New Zealand. Australian responses were confined to diplomatic protests and preventing delivery of the second of the Fiji allocations within the Pacific Patrol Boat programme, but in the policy statements of the New Zealand Government, procurement is explicitly linked with the ability to intervene in the south-west Pacific. The joint production of the new light patrol frigate with Australia is put in the context of the creation of a 'task force' including these ships, a replenishment vessel (bought from South Korea) and a yet-to-be-procured support ship. The task force would be required to meet the criteria of flexibility laid out by Defence Minister Tizzard and in the 1987 Defence White Paper. These were to offer a limited capacity for intervention, but Tizzard stressed as more important the ability to protect New Zealand interests in its 200-nautical-mile EEZ 'and help the Pacific Island countries to do the same'.[53] However, during major disturbances on the island of Vanuatu during May 1988, New Zealand again, like Australia, sent anti-riot gear and medical supplies to the island but no military forces.[54]

What is clear is that the procurement programmes of Australia and New Zealand, together with the Pacific Patrol Boat programme and the associated changes in maritime surveillance operations, are manifestations of the emergence of a regional security system in the South Pacific, where part of the definition of 'self-reliance' stressed by both Australia and New Zealand includes a desire that the military presence of foreign major powers be excluded to the extent possible; and in pursuit of this goal, both have identified a need to take a greater role themselves in promoting regional stability.

Notes and references

1   Mack, A., *Australia's Defence Revolution*, working paper no. 150 (Research School for Pacific Studies, Australian National University: Canberra, 1988), p. 9.

2   *Defence of New Zealand: Review of Defence Policy* (Government Printers: Wellington, 1987), pp. 22–23.

3   For example, whereas in the 1960s roughly 50 per cent of New Zealand exports went to Britain, New Zealand's three largest trading partners are Japan, Australia and the United States.

4   Smith, C., 'Seeking a new role', *Far Eastern Economic Review*, 8 June 1989, pp. 51–55.

5   Lewis Young, P., 'The Australian Navy: facing up to the problems and restructuring for the future', *Navy International*, Feb. 1984, pp. 91–99; Lewis Young, P., 'The Australian FFG-7', *Navy International*, June 1986, pp. 373–76.

6   Lewis Young, P., 'The ANZAC ship project', *Navy International*, Dec. 1987, p. 601.

7   Lewis Young (note 6); 'Williamstown Dockyard sold to AMEC', *Asian Defence Journal*, Feb. 1988, p. 91.

8   'From cornerstone to millstone: Australia's strategic inheritance', *Pacific Defence Reporter*, Jan. 1983, p. 1.

9   Mack (note 1), p. 9.

10   *The Defence of Australia* (Australian Government Printing Service: Canberra, 1987); and *Defence of New Zealand: Review of Defence Policy* (note 2).

11   Dibb, P., *Review of Australia's Defence Capabilities*, Report to the Minister for Defence, Mar. 1986 (Australian Government Printing Service: Canberra, 1986); and Cooksey, R. J., *Review of Australia's Defence Exports and Defence Industry*, Report to the Minister for Defence (Australian Government Printing Service: Canberra, 1986).

12   *Financial Times*, 7 Jan. 1986.

13   Foxwell, D., 'Defence White Papers from Australia and New Zealand', *Military Technology*, June 1987, pp. 68–75.

14   Quoted in Babbage, R., 'Should Australia plan to defend the Christmas and Cocos Islands?', *Canberra Paper on Strategy and Defence*, no. 45 (Australian National University: Canberra, 1988), p. 1. It is also worth noting a heightened Australian interest in Indian naval policy, which has led to a reconsideration of the decision to withdraw defence attaches from India and Pakistan in April 1987: 'Defence rethink', *Far Eastern Economic Review*, 23 June 1988, p. 9.

15   Babbage, R., 'Australia's new defence direction', *Pacific Review*, no. 1, 1988, pp. 93–94.

16   Cranston, F., 'RAAF AWACS programme launched', *Jane's Defence Weekly*, 30 July 1988, pp. 156–57.

17   Babbage, R., 'Coastal surveillance and protection: current problems and options for the future', working paper no. 15 (Australia National University, Canberra, 1988), p. 33.

18   The chief source of information here is *The Defence of Australia* (note 10) and reports of its presentation to the Australian Parliament in Mar. 1987 by Minister for Defence Kim Beazley.

19   Beazley, K., 'Defence Policy Information Paper', Office of the Minister for Defence, Canberra, 19 Mar. 1987, p. 2, quoted in Mack (note 1).

20   *AAS-Milavnews*, June 1987, pp. 2–3.

21 *AAS-Milavnews*, July 1987, pp. 2–4; Hamilton, I., 'Only small Budget increase but White Paper momentum will be maintained', *Pacific Defence Reporter*, Oct. 1988, pp. 22–23.
22 This is discussed more fully under the New Zealand section.
23 Gubb, M., *The Australian Military Response to the Fiji Coup*, working paper 171 (Australian National University, Research School of Pacific Studies, Nov. 1988); Subrahmanyam, K., 'The coup in Fiji: the strategic implications', *Times of India*, 13 June 1987.
24 Done, K., 'Kockums in Australian submarine order', *Financial Times*, 19 June 1987.
25 Grazebrook, A. W., 'The new submarine project', *Pacific Defence Reporter*, July 1987, pp. 35–36.
26 Hamilton, I., 'Lean time ahead for defence, whatever the election outcome', *Pacific Defence Reporter*, July 1987, p. 46.
27 Robertson, F., 'Jackpot for industry', *Pacific Defence Reporter*, July 1987, p. 38.
28 'Canberra threatens to build submarines abroad', *Jane's Defence Weekly*, 23 Apr. 1988, p. 787.
29 Cheeseman, G., 'The Australian arms trade: patterns, policies and prospects', working paper no. 46 (Australian National University, Canberra, 1988), especially pp. 1–7.
30 *Financial Times*, 21 Jan. 1987.
31 Eagles, J. and James, C., 'Barriers fall Down Under', *Far Eastern Economic Review*, 7 July 1988, p. 76–77.
32 'Australia and Sweden in technology talks', *Jane's Defence Weekly*, 5 Sept. 1987, p. 479.
33 *Jane's Defence Weekly*, 30 Apr. 1988; 'ANZAC competition narrows to two', *Naval Forces*, special issue 'The Naval Balance 1988', p. 124.
34 'Williamstown Dockyard sold to AMEC', *Asian Defence Journal*, Feb. 1988, p. 91.
35 *Defence of New Zealand: Review of Defence Policy* (note 2), p. 34.
36 *Defence Review*, Journal of the New Zealand House of Representatives, 1983 (Government Printers: Wellington, 1983).
37 'NZ doubts hit ANZAC ships', *Jane's Defence Weekly*, 3 Sep. 1988, p. 431; Cranston, F. and Launder, I., 'NZ faces MEKO purchase problems', *Jane's Defence Weekly*, 12 Nov. 1988, p. 1179.
38 *Defence of New Zealand: Review of Defence Policy* (note 2),
39 Barber, D., 'Lange unveils new defence strategy', *The Independent*, 27 Feb. 1987.
40 The New Zealand domestic political environment is discussed at length in Burnett, A., *The A-NZ-US Triangle* (Australian National University: Canberra, 1988), especially in chapters 2 and 4.
41 Sections of the report are reproduced in Lewis Young, P., 'Defending New Zealand after ANZUS', *Jane's Defence Weekly*, 13 Feb. 1988, pp. 263–65.
42 Lange, D., Speech to the Wellington Institute of International Affairs, reproduced in *Pacific Islands Monthly*, Apr. 1988, pp. 39–41.
43 Cranston, F., 'ANZUS dispute affects RAN's frigate project', *Jane's Defence Weekly*, 11 Jan. 1987, p. 56.
44 Lewis Young (note 41), pp. 263–65.
45 Barber, D., 'NZ unmoved by loss of US ally status', *Independent*, 26 Sep. 1987.

46   New Zealand Ministry of Defence News Release, 18 Sep. 1987, reproduced in *Navy International,* Dec. 1987, p. 600.

47   Tsamenyi, B. M., 'The exercise of coastal state jurisdiction over EEZ fisheries resources: the South Pacific practice', *Ambio,* vol. 17, no. 4, 1988, pp. 255–58.

48   'South Pacific: defence initiatives', statement to Parliament reproduced in *Australian Foreign Affairs Record,* Feb. 1987, pp. 72–75.

49   Babbage, R., 'Australia and the defence of Papua New Guinea', *Australian Outlook,* Aug. 1987; 'Australia-PNG agreement', *Jane's Defence Weekly,* 19 Dec. 1987; Cranston, F., 'Australia/Papua New Guinea forge closer defence ties', *Jane's Defence Weekly,* 22 Aug. 1987, p. 305.

50   *The Defence of Australia,* Australian Department of Defence White Paper (Government Publishing Service: Canberra, 1987), p. 18; 'Papua New Guinea to expand military', *Defense & Foreign Affairs Daily,* 6 Oct. 1987, p. 2; 'Australia to fund Solomon listening post', *Jane's Defence Weekly,* 7 Nov. 1987, p. 1026; 'Vanuatu: emergency assistance', *Australian Foreign Affairs Record,* Feb. 1987, pp. 75–76.

51   Tsamenyi (note 47), pp. 255–58.

52   Statement by Foreign Minister Hayden, 29 Sep. 1987.

53   New Zealand Ministry of Defence News Release, 18 Sep. 1987, reproduced in *Navy International,* Dec. 1987, p. 600.

54   Fathers, M., 'Foreign forces to aid Vanuatu', *The Independent,* 18 May 1988.

6. Argentina and Brazil

I. Changing security perceptions

In spite of their historical rivalry, there is a great deal of symmetry in recent naval developments in Argentina and Brazil. While Argentina and Brazil retain some specific national problems and perceptions, this symmetry in policy is in part a reflection of a common re-evaluation of changes in the local security environment and of the role of Latin America within the international community. This re-evaluation dates to the period of the late 1970s, when it became clear that the bright economic future predicted for Latin American countries was not about to materialize. As a result, both countries sought to design policies that maximized those advantages which were available. In both countries two basic areas of advantage were identified independently but simultaneously: geographical location and those scientific and technical programmes which had already been under way for some time and so were reasonably advanced.

The principal geographical advantage of Argentina is its proximity to Antarctica and the South Atlantic, areas identified as of significant future economic importance. Through the 1970s a perception emerged that Argentina actully consisted of three constituent regions, the South American mainland territory, Antarctica and an 'island continent' made up of South Atlantic islands and the 200-nautical-mile EEZs radiating out from them.[1] For Brazil, the principal advantage of geography is the existence of enormous domestic natural resources. However, in the 1970s it also appeared that Brazil was in a position to take advantage of an expansion in trade between the Americas and Africa anticipated in the wake of Portuguese decolonization, in particular. Both Argentina and Brazil have sought to take maximum advantage of existing scientific research programmes, including nuclear programmes.[2]

At the end of the 1970s therefore, the future development of Argentine and Brazilian foreign policy rested on two assumptions.

First, they would in future be regional partners seeking to limit extra-regional intrusion rather than rivals drawing in foreign powers to compete with one another. The development of a Latin American regional security system was further stimulated in 1982 by the Falklands/Malvinas War, which crystallized attitudes among local countries that had previously been affected by conflicting currents. In particular, countries with a common Spanish heritage and, to a lesser extent, Brazil were supportive of the Argentine position. This proved more powerful than either Pan-Americanism or deference to the attitude of the United States, which was highly critical of Argentina's resort to force.[3] Second, only through improving the range and level of domestic high technology could they hope to take advantage of economic opportunities available within the region.

These developments have had a direct impact on naval planning. In both Argentina and Brazil naval programmes include plans for a spectrum of new vessels ranging from patrol vessels and corvettes to nuclear-powered submarines. Moreover, in both cases these plans include a major component of domestic construction. However, there are also some significant differences between the experiences and plans of the two countries.

II. Argentina

The influence of the Falklands/Malvinas War

In Argentina the greatest shock to this new perception of foreign policy was the Falklands/Malvinas War of 1982. The origins of the decision to invade and the rights and wrongs of the dispute are not at issue here, only the consequences of defeat for Argentine naval policy. There were basically two effects: a perception that there was little point in relying on international law in a dispute with a major power; and a greater emphasis on maritime operations throughout the South Atlantic rather than in the immediate coastal area.

Although of questionable validity, Argentine attitudes to international law were affected by the perception that Britain had violated the Treaty of Tlatelolco by the deployment of nuclear-powered submarines in the region and through an implied threat of use of nuclear weapons.[4] The emphasis on naval forces capable of maritime operations beyond the immediate coastal environment has principally been visible in the programme to buy and construct under licence a fleet of large conventionally powered submarines designed in the Federal Republic of Germany.

Current Argentine naval programmes

One interesting aspect of the naval programme in Argentina is the exclusion of a certain programme. The aircraft-carrier *Veinticinco de Mayo* is now 43 years old and, although studies regarding a replacement have been made, no decision on a replacement is imminent because of economic shortages.[5] The aircraft-carrier cannot be replaced for financial reasons, and recent modernization has been limited to the replacement of British electronic systems, although more extensive modernization is planned. Super Etendard aircraft bought for the carrier are now mostly shore based, and the ship carries 18 fixed-wing aircraft, a mix of A-4Q Skyhawk strike aircraft and S-2E trackers for reconnaissance, anti-submarine missions and target acquisition. The carrier also carries four helicopters, either Sea Kings or Alouette IIIs. The helicopters have very limited surface-search radars and are used for anti-submarine missions.

Procurement programmes which are under way include the second stage of a programme to acquire 10 escort vessels of MEKO design; the production of up to four TR 1700 submarines under licence from the Federal Republic of Germany to accompany two submarines bought directly from Thyssen Nordseewerke; and the refurbishment of 55 A-4 Skyhawk aircraft with the assistance of Israel and, possibly, South Africa. The modernization undertaken since 1977 means that, in spite of economic problems, Argentina is committed to maintaining a surface fleet of around 10 vessels at least through the 1990s, although some of the MEKO-140 ships currently being built may be sold. The Argentine Navy has expressed a requirement for nuclear-powered submarines, but none of their requests have been acceded to.

Before 1982, Argentine surface ships were traditionally organized in two task forces, one based on the aircraft-carrier *Veinticinco de Mayo*, the other on the battle cruiser *General Belgrano*. The Falklands/Malvinas War has re-shaped Argentine defence planning partly because the *General Belgrano* was sunk, but it also reflects the course of the war. Apparently there will still be two task forces, one based around the carrier and the other for escorting convoys, composed of MEKO-140 frigates and French A-69 frigates bought in the late 1970s. However, these plans are contingent on the retention of the MEKO frigates.

In Argentina, programmes initiated in 1978 and 1980 with the West German company Blohm and Voss were planned to lead to the delivery of 10 escort vessels to the Argentine Navy. The first contract, for the sale of six MEKO-360 destroyers, was signed in 1978. In October 1980, the contract was modified to cover the sale of four destroyers and the production under licence of six of a scaled-down version of the same design, the MEKO-140 frigate, in Argentina. The programme has been amended subsequently, and in October 1987, the Argentine Navy confirmed that the Government was negotiating the sale of the final two MEKO-140 frigates.[6] The entire frigate programme has been jeopardized by the financial problems in Argentina. At one time it appeared that all six ships would be sold as they were completed.[7] The building of the MEKO-140 class has been relatively stable at one ship completed per year, but there are reports that Argentine naval industries may become increasingly focused on the modernization of existing vessels rather than new construction. In 1988, the *Veintecinco de Mayo* began a two-year refit aimed at allowing the operation of Super Etendard aircraft on a regular basis. Meanwhile, submarine construction had been suspended in favour of upgrades of the electronics in West German-built vessels. A continuous problem was also the effort to maintain two Type 42 destroyers of British origin without access to any spare parts from the UK.[8] However, the Argentine Navy is a considerable landowner and industrial power in Argentina and also has other financial interests which the Government has sold to raise funds.[9]

Submarine programmes

A third arm of an Argentine naval triad will now be a submarine arm which is planned to include seven vessels, the production of TR 1700s replacing West German Type 209s bought in the 1970s.[10] In 1977 the Argentine Navy issued a call for bids to supply a new submarine. Among the respondents was the West German design company Thyssen Nordseewerke (TNSW), which began work on a new design specifically to meet the bid.[11] The contract to supply a design subsequently designated the TR 1700 was signed in 1979 and reported to be worth DM 2 billion. The first two submarines were to be built in FR Germany at Emden, and an additional four were to be constructed at the Astillero Domecq Garcia shipyard in Buenos Aires, in which TNSW had a 25 per cent holding.[12] In the event, the final two submarines may never be built for the Argentine Navy. In 1988, work in the Buenos Aires yard had virtually ceased and the West German Government gave permission for the ships to be sold when they are completed, should the Argentine Government so decide.[13]

In 1977, respondents to the Argentine Navy call for bids were quickly presented with a long, detailed and demanding set of specifications for the new submarines. They had to have great endurance, long operational deployment time, high transit speed and large weapon capacity, and required a very short time for battery recharging.[14] This called for a design different from anything available from existing submarine manufacturers. Argentina sought a conventional submarine that incorporated many of the features associated with nuclear-powered submarines. In particular, the boat had to be very big to accommodate a powerful motor, extensive fuel and torpedo storage space. France and Italy both submitted existing designs for the competition, the Daphne and Sauro Classes, respectively. Both were unsuccessful, largely because they could not meet the Argentine operational requirements with the proposed vessel. TNSW were successful largely because, unable to offer the necessary characteristics with existing vessels, they agreed to design a boat specifically for Argentine requirements.[15] In order to construct and maintain the submarines, Argentina has received West German assistance in establishing support facilities including two berths supported by electrical and engineering workshops and an enormous submarine-lifting crane.[16] (These facilities have contributed to the suggestion that Argentina may at some point embark on the construction of much larger, possibly nuclear-powered, submarines.) There have been serious problems with the production of the submarines. The first of these, laid down in October 1983, is tentatively planned to be launched in 1990, with the second in 1991.[17] However, in spite of the slow and disjointed progress on these vessels, a third has apparently now been begun.

Argentina has been linked with the possible production of nuclear-powered submarines in spite of the problems with the conventional submarine construction programme.[18] The possibility has been discussed more seriously in recent years for two reasons: first, because of the increase in nuclear co-operation with Brazil since the visit of Brazilian President Figueiredo to Argentina in 1980 when agreements signed included technical co-operation, joint training and exchange of radioactive material; and second, because in 1983 Argentina announced that it had successfully developed an indigenous capability to enrich uranium (and therefore produce the fuel for a submarine nuclear reactor). Moreover, these developments have taken place against the background of an Argentine nuclear policy to insist on the right to conduct any kind of peaceful nuclear research—in order to 'free Argentina from scientific and technological colonialism'—and a view of itself as the most advanced developing country in terms of nuclear technology.[19] While the ambition to produce an indigenous nuclear-powered submarine may remain, it is clear that there will be no progress towards this in the short term.

Arms embargoes and naval programmes

In 1982 naval contracts placed European governments and, in particular, the governments of France and the FRG in a sensitive position in the context of the Falklands/Malvinas War. France was in the process of delivering Super Etendard fighter-bombers armed with AM-39 Exocet missiles to Argentina when the war broke out. Five had already been delivered and a further nine were on order. Argentina was also an important customer for West German arms exports, accounting for over 35 per cent of exports from the FRG between 1979 and 1983.[20] Outside of the arms relationship, there had been a long history of friendship between the Federal Republic and Argentina. However, the UK was an ally and the MEKO-360 programme was to have included the supply of Lynx helicopters, two of which were delivered by the UK to Argentina prior to the cancellation of the contract in 1982.[21] Moreover, the war occurred towards the end of a detailed reappraisal of arms export policy in the Federal Republic which culminated in the publication of new guidelines for government policy, the *Politischen Grundsätze der Bundesregierung für den Export von Kriegswaffen und sonstigen Rüstungsgütern*, on 28 April 1982.[22]

In April 1982, France, FR Germany and the United States all imposed arms embargoes on Argentina which included all weapons and contracts, including those signed before the imposition of the sanctions on Argentina. In July and August 1982, the UK took a lead in arguing for a removal of EC trade sanctions against Argentina but argued that the arms embargoes should remain. In August 1982, France lifted its embargo on arms sales to Argentina.[23]

FR Germany did not lift its embargo on Argentina at that time, saying that the embargo would remain until the conflict was 'definitively ended in the political sense'.[24] In fact, the West German embargo was lifted in February 1983.

Work continued on both submarine and MEKO construction. The first submarine, laid down in 1980, was not launched until September 1982 and was not scheduled for delivery before 1984. However, the first MEKO-360 was being fitted out when the conflict began, having been launched in March 1981. The embargo probably delayed completion and delivery of the MEKO-360 until it was clear that there would be no immediate resumption of hostilities. The first MEKO-360 was commissioned in February 1983 and was delivered in March 1983.[25]

III. Brazil

Naval policy in Brazil has been characterized by a constant problem of adapting ambitious plans to match scarce resources. The current lack of funding for equipment programmes of the Brazilian Navy contrasts with the performance of a previous 10-year naval plan announced in 1967, which allowed for considerable naval modernization based around a naval construction programme drawing heavily on imported technologies. The funding of these programmes led to Brazil becoming the most modern and capable of the regional Latin American navies by the late 1970s.[26] However, this relative regional advantage has made it difficult for the Navy to argue for a greater share of available funds in the 1980s, especially in the climate of improved relations with Argentina. The Navy has had most success arguing for programmes that support the ability to monitor the offshore estate, on the one hand, and the case for a maritime air component based on helicopters, on the other. The latter programme increases the capacity for offshore surveillance, but also fits with a broader Brazilian aim of developing a manufacturing base for helicopters.

Brazilian naval programmes

In 1984, Brazil published an ambitious 15-year naval programme that envisaged a doubling in the size of the fleet. The centre-piece of the programme was to be the replacement of the ex-Royal Navy 1940s vintage aircraft-carrier *Minas Gerais* sold to Brazil in 1960 and the construction of nuclear-powered submarines in Brazil.[27] All the projects it contained included some form of local construction. British assistance was anticipated in the construction of a second aircraft-carrier, and the nuclear submarine was expected to be a modified version of a design produced by HDW of FR Germany, already engaged as the supplier of Type 209 submarines for the Brazilian Navy. An extensive re-fit programme was envisaged, in which Inhauma class corvettes, originally planned to receive European weapons systems, would be fitted instead with Brazilian equipment. Specifically, Exocet anti-ship missiles would be replaced by a missile called the Barracuda, and Bofors 40-mm guns would be replaced by a point defence system based on an Avibras 20-mm anti-aircraft gun.[28] Moreover, Brazilian naval production has been linked explicitly to developing the capacity to export ships. In 1982 the Government created an organization designed to oversee and speed up the development of naval production, Emgepron (Empresa Gerencial de Projetos Navais), which was linked to the Navy, and the director of naval production announced an annual export goal of $250 m.[29]

The Brazilian Navy has discussed the acquisition of two very different versions of a new aircraft-carrier to replace the *Minas Gerais*. On the one hand, plans for the acquisition of a vessel of 40 000 tonnes intended to carry fighter aircraft were announced in 1983.[30] On the other hand, a more modest vessel of around 25 000 tonnes has also been discussed. The 1983 plan depended on agreement between the Brazilian Air Force and the Brazilian Navy, the former having a monopoly on the operation of fixed-wing aircraft that they have refused to give up. The Air Force has had a monopoly on the operation of fixed-wing aircraft since a January 1965 Presidential Decree ended a dispute over the issue of naval aircraft by limiting the Navy to an all-helicopter force.[31] To some extent, the Falklands/Malvinas War provided a catalyst for a request for funds by the Navy, but the acquisition of carrier-based aircraft would also strengthen the case for an eventual replacement of the existing aircraft-carrier. Aware of the budget problems involved in the expansion of the fleet air arm, the Navy investigated the possibility of buying second-hand carrier-based aircraft, in particular the A-4 Skyhawk. Kuwait and Israel seemed likely sources of aircraft as both of them were in the process of evaluating replacements for the Skyhawk in their own air forces.[32] However, the acquisition of second-hand aircraft would still have had significant consequences. The *Minas Gerais* would have had to be significantly altered to operate A-4s, notably by altering the catapult and arresting gear on board. In 1983, the carrier had only recently emerged from a complete modernization intended to extend its operational life into the 1990s, and an immediate return to the shipyard would have involved additional expenditure. In the event, the plans for a purchase of carrier-based fixed-wing aircraft were cancelled in October 1984, and the Brazilian Navy Minister announced the possibility of a replacement for the *Minas Gerais* in the region of 25 000 tonnes.[33]

There has also been considerable discussion related to the possible production of a nuclear submarine or the adaptation of a large conventional submarine to house a nuclear propulsion system of Brazilian design by the year 2000. According to the Navy Minister, around $38 million has already been spent on the project, and several countries have been approached with regard to design and technology transfer.[34] Other accounts suggest that one of the functions of the Aramar nuclear facility west of São Paulo is to develop a nuclear reactor suitable for a submarine using uranium enriched in Brazil.[35] Brazil apparently plans a progression from the production of the West German Type 209 currently under construction for the Brazilian Navy to the manufacture of relatively large conventionally powered submarines to a Brazilian design (although heavily derivative of the Type 209), known as the NAC-1. This would in turn lead into the development of a nuclear submarine of 2700–3000 tonnes.[36] The reactor itself is apparently already operational in Brazil. However, the earliest date for any possible submarine commission is 2010.[37]

The impact of economic shortages

In reality, in spite of the scale of these plans, there is no evidence of any new programmes in being since 1984, other than the building of two lighter classes of patrol boat of less than 400 tonnes in displacement. It is clear that there is a significant mismatch between public statements of intent by the Brazilian Navy and real-world procurement programmes.

Public statements have often been at odds with the actual progress of procurement programmes. For example, in October 1981, the Brazilian Government signed a contract for the construction of 12 Inhauma class corvettes in Brazilian shipyards, according to designs produced by the West German company Marine Technik. In the middle of 1986, the programme was expanded to include the eventual construction of 16 vessels. However, the first corvette was not launched until March 1987 and delays in the programme have meant that a maximum of 4 can now be completed by 1990. In the original order, all 12 corvettes were planned for completion by this date.[38] Moreover, the original plan included the gradual replacement of French surface-to-surface and surface-to-air missile systems with weapons of Brazilian manufacture. In 1988, the programme intended to develop a naval surface-to-surface missile, the Barracuda, has been frozen owing to a lack of funds. The programme intended to produce a naval point defence system seems to have been cancelled altogether.[39]

None of the programmes under way have progressed smoothly, and some have not progressed at all. From the late 1970s, Brazil discussed the purchase of submarines to replace Guppy class boats built during the 1940s and sold to Brazil by the United States in the 1970s.[40] In 1982, Brazil had apparently agreed to purchase submarines from the Federal Republic of Germany. The contract seemed to cover the sale of two submarines. The first would be built in Kiel, but with the close participation of an on-site Brazilian technical team, and the second would be assembled in Brazil. The contract followed months of discussion initiated in the aftermath of the Falklands/Malvinas War.[41] However, there was a two-year delay in the conclusion of a contract because of Brazil's difficulties in financing the purchase. In February 1984, a contract was finally signed with the West German firm HDW for the acquisition of new submarines, the Type 209, displacing 1400 tonnes and of a design similar to that of vessels in service with Argentina, Colombia, Peru and Venezuela. The contract included an option on two further boats, part of which may have been exercised. The first of these boats was to be built in Kiel; at least two were subsequently to be built at the Arsenal de Marinha in Rio de Janeiro. The first vessel from West German production was completed in 1988, and delivered in the second half of the year. Production in

Brazil has been delayed, although there were reports that the two submarines would be laid down simultaneously in 1988.[42]

As noted above, some naval procurement programmes in Brazil have been affected by inter-service rivalries, but all of the naval programmes planned have been affected by financial shortages. (Brazilian military expenditure has been consistently less than 1 per cent of gross domestic product.[43]) During the 1980s, Brazilian officials have been investigating a series of possible arrangements to fund naval procurement programmes. In the protracted negotiations with HDW over the supply of submarines in 1983, Brazil investigated the possible shipment of iron ore to FR Germany in part payment.[44] To meet the cost of the construction of eight large offshore patrol vessels, the Brazilian Navy has been allocated a percentage of revenues from offshore oil production.

Brazil has also relied on slippage—deliberate delay of programmes into subsequent budget years—and a search for second-hand vessels in order to fulfil its naval programme.[45] In the early 1980s, Brazil initiated a search for second-hand ships in reasonable repair. It was hoped to buy three Charles F. Adams frigates retired by the West German Navy. However, the Federal Republic modernized these vessels between 1983 and 1986, and they will now serve into the 1990s. On being approached, the United States has offered older vessels for sale or lease.[46]

Exploiting maritime economic resources

The progress of naval procurement programmes in Brazil is illustrative of the primacy within Brazilian maritime policy of issues relating to the the 200-nautical-mile sea limit declared in 1970. As noted above, Brazilian naval procurement is in reality largely restricted to programmes already under way by 1984, other than in the area of the construction of offshore patrol vessels for coast guard duties and the surveillance and policing of Brazil's Exclusive Economic Zone. The funding of a class of eight offshore patrol vessels has been guaranteed by providing a percentage of oil revenue. These vessels, in the region of 100 metres in length and 1100 tonnes in weight, will be armed by stripping some of the armament from 1940s-vintage Fletcher class destroyers, notably 5-inch guns and 40-mm guns. Construction of the first of these vessels was due to begin in 1987, with delivery scheduled for 1989.[47]

Brazil's concern with the issue of exploiting offshore assets was demonstrated as early as 1971, when Brazil inserted a caveat on signing the Treaty on the Prohibition of the Emplacement of Nuclear Weapons and Other Weapons of Mass Destruction on the Seabed to the effect that 'nothing in the treaty shall be interpreted as prejudicing in any way the sovereign rights of Brazil in the area of the sea, the seabed and the subsoil thereof in the area adjacent to its coasts'.[48]

The trend within Brazilian naval procurement policy suggests a movement away from a fleet based on major surface combatants through a gradual process of replacing old and relatively heavily armed capital ships of large tonnage with a mixture of less heavy corvettes and offshore patrol vessels. Resources for new construction of capital ships have been committed to the Type 209 submarine programme, leaving little for the construction of new classes of surface combatants. The impression that Brazilian naval procurement is less concerned with force projection or operations beyond territorial sea space is reinforced by the growth of co-operative maritime relationships in a number of areas. Whereas on the Western seaboard of Latin America there have been some bitter disputes over the demarcation of national maritime jurisdiction (notably between Argentina and Chile, between Peru and Ecuador and between Chile and Peru/Bolivia), Brazil's relationship with its neighbours over maritime security issues has been marked by stability and progress. During the 1980s, Brazil and Argentina have carried out a number of joint naval exercises and investigated joint naval programmes and production, including, as noted, exploring the possibilities of collaboration in such highly sensitive areas as nuclear technology. During the United Nations Conference on the Law of the Sea, Brazil formed a number of loose coalitions with other Latin American countries in defence of its interests, notably with regard to issues of EEZs and exploiting economic resources on the sea bed.[49]

Notes and references

1   Wixler, K. E., 'Argentina's geopolitics and her revolutionary diesel-electric submarines', *Naval War College Review*, winter 1989, pp. 86–108.
2   For an overview of nuclear policy in Brazil and Argentina see Goldblat, J. (ed.), SIPRI, *Non-Proliferation: The Why and the Wherefore* (Taylor & Francis: London and Philadelphia, 1985).

3   On the other hand, the Falklands/Malvinas War isolated those Latin American countries, notably Chile, which did not support Argentina and generated further tension between Argentina and Chile; Child, J., The impact of the Falklands/Malvinas conflict on the inter-American system, paper delivered to the International Studies Association, 5 Nov. 1982.

4   The British Ministry of Defence refused to confirm or deny that the Royal Fleet Auxilliary *Fort Austin* carried nuclear depth charges during the Falklands/Malvinas War: Fursdon, E., 'Silence on nuclear dummy', *Daily Telegraph*, 3 Nov. 1982. For a discussion of the legal aspects see Goldblat, J. and Millan, V., 'The Falklands/Malvinas Conflict', *World Armaments and Disarmament: SIPRI Yearbook 1985* (Taylor & Francis: London, 1985), pp. 467–95.

5   *Jane's Fighting Ships, 1988–89* (Jane's: Coulsden, 1988), p. 10.

6   Peru and Ecuador were two potential customers, along with Portugal, *Jane's Defence Weekly*, 17 Sep. 1987.

7   *Proceedings of the US Naval Institute*, Mar. 1988, p. 28.

8   'Latin American naval developments', *Navy International*, Sep. 1988, p. 413; 'Sub construction/Refit in Argentina', *Naval Forces*, Oct. 1988, p. 93; 'Argentine ship update now under way', *Jane's Defence Weekly*, 22 Oct. 1988, p. 999; 'Argentina upgrades submarines', *International Defense Review*, Nov. 1988, p. 1523.

9   'The Military Inc.: the Armed Forces in business', *South*, Mar. 1988, p. 13; *Proceedings of the US Naval Institute*, Mar. 1988, p. 28.

10   Murguizur, J. C., 'The future of the submarine in Argentine naval policy', *International Defense Review*, Apr. 1984, p. 453.

11   *Naval Forces* Special Supplement, no. 3, 1985, p. 22

12   Murguizur (note 10), p. 454.

13   *Jane's Fighting Ships 1988–89* (note 5).

14   Wixler, K. E., 'Argentina's geopolitics and her revolutionary diesel-electric submarines', *Naval War College Review*, winter 1989, pp. 86–108.

15   Dicker, R. J. L., 'The German submarine industry: will the success continue?', *International Defense Review*, Sep. 1983, p. 1178.

16   Dicker (note 15), p. 1293.

17   *Jane's Fighting Ships, 1988–89* (note 5).

18   English, A. J., 'Argentina to build SSN', *Defence*, Jan. 1989, p. 8.

19   Pilat, J. F. and Donnelly, W. H., *An Analysis of Argentine Statements on the Purpose and Direction of Argentina's Nuclear Policy*, Report for the Congressional Research Service, 22 Apr. 1982.

20   Anthony, I., 'The trade in major conventional weapons', *SIPRI Yearbook 1989: World Armaments and Disarmament* (Oxford University Press: Oxford, 1989).

21   *Jane's Fighting Ships 1988–89* (note 5), p. 12; *AAS-Milavnews*, Jan. 1983, p. 1; *Defense & Foreign Affairs*, 14 Feb. 1983, p. 1.

22   Courades Allebeck, A., 'Arms trade regulations', *SIPRI Yearbook 1989* (note 20).

23   Freedman, L., 'The war of the Falkland Islands 1982', *Foreign Affairs*, fall 1982, p. 200; 'Argentina to rearm; with priority to air force', *Washington Times*, 27 Aug. 1982, p. 6; Osborn, A., 'EEC moves to end trade sanctions against Argentina', *Daily Telegraph*, 26 Oct. 1982.

24   'West Germany: no arms to Argentina', *Defense & Foreign Affairs Daily*, 17 Aug. 1982, p. 1.

25   'Argentinian MEKO corvettes', *Naval Forces*, Mar. 1983, p. 87; Speck, M., 'Argentina gets new destroyer', *Daily Telegraph*, 20 July 1983; Argentina received first new destroyer', *International Herald Tribune*, 23 Mar. 1983.

26   For a discussion of regional naval power, see Morris , M. A., *The Expansion of Third World Navies* (Macmillan: London, 1987), especially chapters 5 and 6.

27   Whitley A., 'Brazil outlines plans for naval re-equipment', *Financial Times*, 12 Sep. 1984; Olive, R. S., '100% fleet increase for Brazilian Navy', *Jane's Defence Weekly*, 29 June 1985, p. 1277.

28   *Navy International*, Aug. 1986, pp. 469–70.

29   Morris (note 26), p. 207.

30   Max, A., 'New attack carrier in Brazil's naval plans', *Jane's Defence Weekly*, 29 Sep. 1984, p. 530; English, A. J., 'Latin American Navies in 1988', *Naval Forces*, Mar. 1988, p. 116.

31   In 1987 Brazilian naval aviation was reorganized into two separate formations, one dedicated to support the aircraft carrier, the other to support the Brazilian Marine Corps: 'Brazilian Navy forms new naval aviation unit', *Jane's Defence Weekly*, 29 Aug. 1987, p. 360.

32   Olive, R. S., 'Brazil seeks Skyhawks for new-role carrier', *Jane's Defence Weekly*, 25 Feb. 1984, p. 274; 'Brazilian Navy evaluates buy of A-4s for carrier deployment', *Aviation Week & Space Technology*, 25 June 1984, p. 229.

33   'No naval strike aircraft', *AAS-Milavnews*, Oct. 1984, p. 5.

34   'Brazilian nuclear submarine plans reported', *Defense & Foreign Affair Daily*, 1 May 1987, p. 4; 'Brazil planning nuclear submarine fleet', *Jane's Defence Weekly*, 19 Dec. 1987, p. 1445.

35   'Brazil's SSN: more details', *Jane's Defence Weekly*, 30 Apr. 1988, p. 837; House, R., 'Brazil says uranium enriched', *Washington Post*, 10 Sep. 1987, p. 29.

36   Meason, J. E., 'Brazil's plans to build a nuclear submarine industry', *Jane's Defence Weekly*, 23 July 1988, pp. 139–40.

37   Scheina, R. L., 'Latin American Navies', *Proceedings of the US Naval Institute*, Mar. 1989, p. 124–29.

38   'Brazil launches Inhauma, first of new corvette class', *Jane's Defence Weekly*, 21 Mar. 1987, p. 490.

39   'Brazil drops weapons in budget response', *Jane's Defence Weekly*, 23 July 1988, p. 113.

40   'Brazil: submarines from FRG', *Defense & Foreign Affairs Daily*, 14 Dec. 1982.

41   'Brazil to buy W. German submarines', *Financial Times*, 6 Aug. 1982.

42   *Jane's Fighting Ships 1988–89* (note 5), p. 51; 'Tupi, Brazil's first Type 1400 submarine', *Jane's Defence Weekly*, 13 Feb. 1988, p. 255.

43   Brazilian defence expenditure has been difficult to estimate accurately because of the tendency to hide defence spending in other headings within the budget and because of a domestic inflation rate of over 350 per cent per year: Tullberg, R. and Hagmeyer-Gaverus, G., 'SIPRI military expenditure data', *SIPRI Yearbook 1988: World Armaments and Disarmament* (Oxford University Press: Oxford, 1988), pp. 151, 172.

44   *Financial Times*, 12 Sep. 1984; McLeavy, R., 'Brazil to barter iron ore for submarines', *Jane's Defence Weekly*, 17 Mar. 1984, p. 394.

45   *O Globo* (Rio de Janeiro), 13 Oct. 1986.

46   Scheina (note 37), p. 29.

47 *Navy International*, Aug. 1986, pp. 469–70.
48 The treaty is reproduced in Goldblat, J., 'The Seabed Treaty', *Ocean Yearbook 1* (University of Chicago Press: Chicago, 1978), pp. 402–11.
49 Brazilian maritime policy is discussed in Morris, M. A., *International Politics and the Law of the Sea: The Case of Brazil* (Westview Press: Boulder, Colo., 1979).

7. Canada

I. The 1987 Defence White Paper

In June 1987, the Canadian Government published *Challenge and Commitment: A Defence Policy for Canada,* its first Defence White Paper for 16 years.[1] As noted below, the White Paper announced a series of procurement programmes to be paid for through a significant increase in defence spending. Most surprising to many observers of Canadian defence policy was the assessment of the international environment of the 1980s on which the appeal for new equipment programmes was founded. The assessment of Soviet policies contained in the document noted:

It is a fact, not a matter of interpretation, that the West is faced with an ideological, political and economic adversary whose explicit long-term aim is to mould the world in its own image. That adversary has at its disposal massive military forces and a proven willingness to use force, both at home and abroad, to achieve political objectives. Perhaps this is a reflection of a deep-rooted obsession with security, well-founded on the bitter lessons of Russian history. It cannot but make everyone else feel decidedly insecure.

This emphasis on ideological struggle was different in tone from previous statements on defence. The annual assessment of Canadian defence in 1986 had actually made no reference to the Soviet Union at all.

In part the change in emphasis reflected the fact that many of the programmes outlined in the White Paper were part of a wider review process initiated in a very different political climate from that prevailing in 1987. The emphasis on the Soviet threat also reflected the perception in the Department of National Defence (DND) that there was a growing gap between Canada's defence commitments and defence resources. The problem for the Government was how to win public support for a significant increase in expenditure.

The 1987 White Paper was the outcome of a more long-term process of decision-making. The document followed a defence review undertaken by the incoming Progressive Conservative Party led by Brian Mulroney after winning an election in 1984. The review process produced two findings. First, Canada was approaching a point where it would be unable to meet existing defence commitments. The Canadian Government had, by 1985, reached a point where Canada's Ambassador to NATO wrote: 'before long the government will have to address hard questions about the size of the defence budget and the manpower ceiling . . . for the choice will be between spending more money or reducing commitments'.[2] This view had been consistently advanced by members of the Sub-committee on National Defence of the Standing Senate Committee on Foreign Affairs. In January 1982 and May 1983, reports of the Sub-committee produced recommendations that were subsequently to be reflected closely in the defence policy of the Mulroney Government.[3] The work of the Sub-committee also helped to shape international perceptions of Canada's role in NATO.[4]

The second finding was that there were new requirements for Canada's defence arising out of the increase in Soviet activities in the north-east Pacific and Arctic regions. As late as 1986, the annual review of the Canadian Department of National Defence defined Canada's direction of defence as being 'to retain and strengthen Canada's military presence in Europe and to continue with the collective defence arrangements that Canada has with the North Atlantic Treaty Organization and the North American Aerospace Defence Command'.[5] One year later the official position of the DND was that there was a need for a significant and visible increase in the overall effectiveness of Canadian forces to allow the Navy to conduct operations in the Atlantic, Pacific and the Arctic.[6]

In consequence of this change in perspective, much of the Defence White Paper was devoted to describing the procurement policy and programmes needed in support of the opening assessment of Canadian defence policy in the White Paper. The White Paper confirmed that in the Government's view Canada was unable to meet its existing commitments, and promised to 'restore the credibility of Canada's contribution to the Alliance through . . . steady, predictable funding'.[7] Moreover, the Government also addressed the cost of the programmes and the necessity for increased defence expenditure in Canada. Defence Minister Perrin Beatty suggested paying for the programmes outlined in the White Paper with an increase in Canadian defence spending of 2–2.5 per cent a year in real terms sustained for 15 years.[8] Finance Minister Michael Wilson apparently guaranteed this minimum increase without imposing any ceiling on expenditure. In fact, during the annual debate on the budget in the Canadian House of Commons, Wilson announced an increase of 6.2 per cent in defence expenditure for the year beginning 1 April 1988, although the figure eventually agreed represented a 2.7 per cent increase.[9]

II. Current naval programmes in Canada[10]

The major programmes currently either under way in Canada or outlined in the White Paper that centrally concern the Canadian Maritime Command are as follows. First and foremost, the White Paper contained a plan to modify the Canadian Submarine Acquisition Program (CASAP). CASAP had focused on the need for conventional submarines to replace three Oberon class submarines bought from the United Kingdom between 1965 and 1968. The new plan envisaged the acquisition of up to 12 nuclear-powered attack submarines which, although based on foreign designs and incorporating many foreign components, would be produced in Canada. The 1989 Canadian budget announced in Parliament at the end of April included a reduction in the defence appropriation that has halted the immediate progress of the nuclear submarine programme and postponed decisions on other programmes.[11] It is not clear at the time of writing whether the programme has been permanently removed from the defence plans of the Canadian Government. The submarine programme involved more than the acquisition of submarines themselves but was part of a far-reaching reconsideration of Canada's submarine environment. Moreover, recent actions and statements by the Government are ambiguous on the future of the programme. In February 1989, the reshuffle of the Canadian cabinet moved two of the strongest supporters of the programme from their posts. Defence Minister Perrin Beatty became Minister of National Health and Welfare, while Associate Minister of National Defence Paul Dick moved to the Ministry of Supply and Services. This led to immediate reports that the submarine programme was under pressure.[12] However, at the time of the ministerial changes and as late as March 1989, Prime Minister Mulroney told a press conference that cancelling the programme was not a viable option. Instead the programme may be spread over 30 years instead of the planned 15–20.[13] The new defence minister William McKnight further complicated the assessment when he announced that the naval staff were preparing alternatives to the nuclear-powered attack submarine (SSN) programme. McKnight said: 'SSNs were the best vehicle for the task. We will now proceed to the second best'.[14] The submarine programme is discussed in a special section below.

The central programme in Canada's naval procurement is the production of six new patrol frigates of the City class under a Ship Replacement Programme (SRP), first outlined in 1977, which led to the laying down of the first frigate in 1987. Other naval shipbuilding programmes include the update and modernization of Tribal class destroyers and the NATO frigate programme. Finally, plans have been mooted for the procurement of a new force of mine countermeasure (MCM) vessels. The MCM programme is a long-term programme likely to begin with the lease of vessels for training purposes before any decision to buy or build minehunters and minesweepers in Canada.[15]

The City class programme is the largest contract ever placed by the Canadian Government with a domestic company, although it is surpassed in cost by the New Fighter Aircraft Programme. This purchase, also outlined in 1977, led to an order for the CF-18 (a version of the US F/A-18 Hornet) in April 1980. By the end of 1988, 138 of these aircraft had been delivered to the Canadian Air Force although 6 of these had subsequently been lost in training accidents. If the CF-18 is not a central element of naval procurement, it is worth noting in this context, since of the 60 aircraft retained in Canada (the rest being assigned to Europe) some are assigned to the attack role.[16]

The Defence White Paper included other maritime aircraft programmes. Linked to the City class construction and the modernization of other vessels was a decision to purchase up to 50 new shipborne helicopters. In addition, the improvement of Canadian maritime reconnaissance and patrol capabilities with the modernization of existing CP-121 Tracker aircraft was planned. Among the ideas suggested in this context was the production of Canadian versions of the US P-3C Orion aircraft.

In its tone and its political/symbolic importance, the White Paper has been presented by the Government as a radical departure from previous policy. Presenting the White Paper to Parliament, Defence Minister Beatty stated: 'we must either engage in a major rebuilding of our navy or give up even the pretence of being able to protect our waters . . . The real question is whether Canada can afford to have a modern navy, or perhaps more accurately, whether a three-ocean nation as dependent on trade as Canada is, can afford not to have a navy'.[17] From the point of view of naval procurement, however, the White Paper can be seen as part of an evolutionary process that can be traced to the 1970s. A chronology of important decisions concerning current Canadian naval developments between the 1971 Defence White Paper and the coming to power of the Mulroney Government is as follows. In May 1972 there was a Maritime Policy Review; in 1975, a Defence Structure Review; in 1982–83, a cabinet decision on the size of Canada's future requirement for frigates was revealed during testimony to the Senate Sub-Committee on National Defence. Collectively these developments shaped many of the programmes currently under way.

Of the procurement programmes outlined above, perhaps only the MCM programme represents a truly 'new' procurement programme, since Canada currently has no mine countermeasure ships. The intention is to purchase vessels to be operated by the naval reserve forces. In peace-time these ships would be used for coastal sur-veillance and training missions.[18] However, while all the other programmes were in train before June 1987, the White Paper changed the scope of the CASAP programme dramatically by announcing a decision to purchase nuclear- as opposed to conventionally powered submarines. The Tribal Class Update and Modernization Program (TRUMP), formally initiated in 1986, was actually put together in large part by the Trudeau Government. Canada's destroyer fleet consists of ships built in the early 1970s as anti-submarine warfare (ASW) vessels. The TRUMP programme is intended to convert these ships to an air defence role. While the anti-submarine armament is not to be removed, the centre-piece of the modernization package includes the replacement of the Sea Sparrow air defence missile system with a Mk 41 vertical launch missile system armed with the Standard SM/2 missile to be bought from the United States. The Italian Oto Melara 127-mm gun on the existing vessels will be replaced with a 76-mm gun from the same company and a Phalanx close-in weapon system (CIWS) bought from the United States.[19] The radars on board vessels will also be changed to the SMART radar system (Signaal Multibeam Acquisition Radar for Targeting) produced by Hollandse Signaalapparaten of the Netherlands, a three-dimensional surveillance and targeting radar designed to track small targets at wave height.[20]

The conversion of Tribal class destroyers to an air defence role was planned to take place alongside a new construction programme called the Ship Replacement Program (SRP), first outlined in December 1977, which would lead to the construction of a class of ASW vessels.[21] A formal announcement of intent to build six new ships came in 1978. By 1982 a cabinet decision to expand the size of Canada's requirement for frigates had been taken, and was revealed during testimony to the Senate Sub-Committee on National Defence in June 1983. An invitation for tenders was issued in August 1983.[22] This programme originally envisaged the production of 18 new ASW frigates in three batches of 6. Called the City class, the first of 6 frigates began production in March 1987 and was launched in January 1988. The SRP was planned to have three phases, however a decision to follow the purchase of a second batch of 6 City class frigates with a third and final 6 was subsequently cancelled and the 1987 White Paper discusses a surface fleet of 16 ships, 12 of them ASW frigates and 4 modernized destroyers (possibly to be replaced by the subsequent addition of the NFR-90 NATO frigate).[23] The first batch of City class frigates is called the Halifax class, and *HMCS Halifax* was launched in January 1988. The second batch, the Montreal class, will differ from the first in that it is planned to be longer, to accommodate a different sonar and an additional vertical launch missile system.[24] However, both batches of ship will incorporate large numbers of imported systems. They are to be armed with RGM-84A Harpoon anti-ship missiles manufactured by McDonnell Douglas and torpedoes produced by Honeywell, both of the United States. The first batch is planned to have a Swedish Bofors 57-mm gun, the second an Italian Oto Melara 76-mm gun.[25]

The Canadian Submarine Acquisition Program

The primary mission of the City class ships has been stated to be anti-submarine warfare and the 1987 White Paper is dominated by programmes that indicate new approaches to ASW in Canadian defence policy, most dramatically the planned acquisition of 10–12 nuclear-powered attack submarines. The immediate prospects of Canada buying SSNs appear to be remote, and the long-term future of the programme is not clear. Within the Department of National

Defence the idea of buying a nuclear-powered submarine has been discussed for over 30 years without any decision being taken. Moreover, the suspension of plans for submarine procurement has been the norm in Canada. Specific plans to modernize Canada's submarine force (known as the Canadian Submarine Acquisition Program or CASAP) under discussion during the early 1980s, led to the announcement in 1985 of an intention to buy up to 12 diesel submarines, which attracted bids from six West European countries.[26] This programme was then deferred to 1990 as a result of shortage of funds.[27] Therefore the announcement that 10–12 nuclear-powered submarines would be built was dramatic but was also greeted with a degree of circumspection. In spite of its statement of intent to purchase nuclear-powered submarines, the Canadian Government at no point committed itself to do so. In 1988–89, a project definition programme worth US $23 million was funded, but the earliest planned date that any specific proposal would have been submitted for cabinet discussion was 1990.[28]

The submarine programme was originally planned to unfold over a minimum of 15–20 years, and therefore existing defence policies would have had to survive several general elections. Moreover, the Finance Department made its commitment on defence expenditure subject to an annual review, and Wilson noted that defence-related capital expenditure had to be conducted 'in a manner consistent with fiscal principles'.[29] Given these uncertainties and considering the scale of the changes implied by becoming a nuclear navy it is reasonable to ask how the decision to buy an SSN fleet has emerged. It can be looked at from a number of perspectives.

First, it can be seen in the context of Canada's relationship with NATO—in itself a question with several dimensions. Second, it can be seen in the context of Canada's relationship with the US; and third, it can be seen from the point of view of Canada's interest in defining its sovereignty over territorial waters.

A Canadian SSN force would create no insoluble operational problems within NATO. The Commander of the Maritime Command of the Canadian Armed Forces, Vice Admiral Thomas, has stated that the submarines were not seen only in the context of patrols under the ice-cap in the Arctic Ocean, while Vice Admiral Anderson, the SSN programme manager, has stated that Canada could only envisage actions which would be undertaken in the context of NATO joint operations.[30] The US Navy has described itself as 'the water manager for the allies in submarine operations' in wartime, and the very least that could be imagined is that there would be no mutual interference between patrols by allied submarines.[31] There is no Canadian effort to 'go it alone'. However, there is a question surrounding whether the submarine programme is the most rational way for Canada to use its available resources. The same investment could have been allocated to an alternative mix of conventional submarines, naval aircraft or surface forces. Here too there is a question of whether Canadian SSN operations would complicate patrols by other allies in the Arctic region during peace-time, when allied co-ordination may not function at the same level as could be expected in war.

The choice of the submarine programme under these circumstances suggests that there was a powerful political influence over the decision. The related questions of US-Canadian relations and perceptions of Canadian national interest were paramount in promoting the submarine programme.

It has been difficult for Canada to define its relationship with the United States in the face of the reality of greatly unequal power between the two as measured by any static indicator. In terms of the Canadian domestic political debate, the negotiation, signing and ratification of the US–Canada Free Trade Agreement in 1988 indirectly heightened public consciousness on issues of Canadian sovereignty by focusing public attention on the extent to which US companies have become a dominant force in the Canadian economy.

Contentious issues in the US–Canada relationship include the controversy surrounding the use of Arctic waters claimed by Canada as sovereign territory. In January 1988 Canada and the US signed an agreement on co-operation in the Arctic region which did not resolve the legal status of the north-west passage. The agreement stated that US surface vessels would seek permission before using the waterway (in reality a series of navigable channels connecting the Atlantic and Pacific oceans around the top of Canada). The transit of submarines was not mentioned, and the agreement did not concede US recognition of Canadian sovereignty over some northern waters claimed by Ottawa.[32]

Canadians have voiced concern that the US Navy can currently operate in Canadian national waters that are inaccessible to the Canadian Navy. There is incomplete knowledge in Canada about the pattern of foreign submarine operations in the region, and Canada is largely dependent for information on the United States. Moreover, there have been Canadian suspicions that the number and level of foreign submarine operations under the ice-cap are growing.[33] The planned use of the submarines under the Arctic ice-cap is part of a wider programme which will require an expansion of information gathering by Canadian forces through several means. First, from the construction of a sea-bed sonar network, second from a comprehensive sea-bed survey, and third from towed array sonars trailed from fixed-wing aircraft and helicopters. The tasks of constructing a network of sea-bed sonars and compiling comprehensive charts of the Arctic sea bed is to lead to the construction of the biggest ice-breaking ship ever built, the *Polar 8*, of close to 50 000 tonnes in weight.[34] Defence Minister Beatty acknowledged the scope of the planned naval procurement programme in his speech introducing the White Paper in the Canadian Parliament. He noted: 'the most effective way of conducting anti-submarine warfare operations in the Pacific and Atlantic Oceans would be with a balanced Maritime Force comprising surface vessels with helicopters, nuclear-powered submarines and maritime patrol aircraft . . . Such a balanced fleet will also offer Canada the option of submarine operations under the ice of her northern waters'.[35]

Another contentious matter which further reinforced Canada's desire to demonstrate its national sovereignty was the issue of US control over the submarine programme.

Much attention was focused on whether or not the US would allow British companies to supply the Canadian Navy with submarines if this might weaken the US case in resolving the status of Arctic waters. The US had two possible means to block a UK–Canadian programme. Under a 1958 agreement, the US and the UK must agree before either transfers nuclear technology to a third party; and under a 1959 US–Canada agreement, Congress has a veto over the transfer of US nuclear technology to Canada.[36] The United States could also have indirectly hindered the sale of a submarine to Canada by France—the only other possible supplier—because the Canadian Department of National Defense stipulated that the submarine would have to have a weapons fit compatible with the existing naval inventory. Consequently, the programme would involve the integration of US weapons—Mk 48 torpedoes and UGM-84 Harpoon anti-ship missiles—into the vessel.[37]

In the event, the United States did not seek to block British applications, although the State Department announcement clearly did not imply an endorsement of the decision to buy nuclear-powered submarines. The statement noted that 'the President has determined that if the Canadians select the Trafalgar design, the interests of the United States are best served by agreeing to the British request'.[38] The US assent was not without conditions. Canada would not be allowed access to any British design improvements in the reactor after the agreed purchase date without a separate application for US approval. Moreover, although Canada has large reserves of raw uranium, this would need to be enriched in order to be usable as fuel for a submarine reactor. The Canadian nuclear industry is largely a user of natural uranium and has little enrichment capacity.[39] Consequently Canada planned to ship raw uranium to the United States for enrichment. The US refused to give Canada more than a five-year enriching contract and said that it would take five years to return uranium shipped from Canada in a usable form.[40] In this way too, the US would have sustained considerable control over the Canadian programme.

Whatever the immediate future of the submarine programme, the Canadian Government has gone a considerable way towards exploring the feasibility of acquiring SSNs. Moreover, there are some reasons to believe that the programme will reappear, albeit in a less ambitious framework. Indigenous research programmes into the development of submarine reactor technology appear to be continuing, maintaining the potential for a Canadian-developed SSN to be built at some future time. Two different programmes currently appear to be exploring a hybrid solution to submarine propulsion, combining nuclear and conventional elements. In this system a nuclear reactor would not propel the submarine but would be used to charge batteries that would in turn generate electricity. One programme, named Stratas after the company that is developing it, is apparently well advanced in developing a first prototype reactor. The other, named the Autonomous Marine Power Source (AMPS), is being developed by ECS of Canada.[41] Both reactors would run using enriched uranium fuel and would apparently be available for export, although the President of Stratas underlined that the programme was subject to scrutiny by the External Affairs Ministry and that only Non-Proliferation Treaty signatories would be acceptable clients.[42]

III. Development of Canada's defence industrial base

The SSN programme can also be seen in the context of the discussion in Canada of the role of the defence industrial base within the wider economy. The debate over the programme emphasized that the vessels would introduce new naval shipbuilding and engineering skills into Canada (or restore skills that had existed prior to the mid-1970s), as well as allowing Canadians access to the technologies involved in the propulsion system and sensors. Defence Minister Perrin Beatty claimed that the vessels would ultimately contain a minimum of 65–70 per cent Canadian input.[43] At the root of this pressure to import was the political judgement that Canada must have an indigenous naval shipbuilding capacity.

The necessity to import designs and technology stemmed from a judgement that if Canada was to sustain its shipbuilding capacity it would be necessary to use government contracts to support industrial development. Procurement policy could assist in sustaining several shipyards (creating the option of competitive tendering for government contracts) and spreading work between yards. However, while naval shipbuilding in Canada had occurred cyclically in the past, there were periods in the late 1960s and late 1970s where there were no ships under construction at all. This discontinuity in government ship orders was identified as the principal obstacle to defence industrial planning in the private sector. According to the Senate Subcommittee, it meant

not only that Canada was missing an entire generation of ships; but also that the necessary engineers and project managers have not been developed and retained inside the services . . . Private industry would find it easier to cope if it were not asked to skip whole generations of military technology. Either way, warship design would likely take a great deal less time, and cost far less, if it did not have to be relearned at intervals of fifteen years.[44]

Reacquiring lost capabilities had to involve linkages with foreign suppliers.

These defence industrial considerations very quickly had powerful consequences in Canada which may have lasting implications whatever the future of the SSN programme. Several multinational joint venture companies were formed to compete in the project-definition stage of the programme. The 1987 White Paper was released on the final day of the ARMX 1987 defence exhibition in Ottawa, and on the same day, at the exhibition, the team of Fenco Engineers Inc., Litton Systems Canada and Halifax-Dartmouth Industries was formed. Other teams are headed by CSE Submarine Group Inc. (itself formed from Canadian Shipbuilding and Engineering Ltd and Rockwell International of Canada); Marine Industries Ltd (MIL) and the SNC Group.[45] These Canadian teams planned to link themselves with overseas contractors from the UK (Vickers Shipbuilding and Engineering Ltd) or France. Vickers offered the Trafalgar class, the latest SSN being built for the Royal Navy, and the French DCN dockyard in Cherbourg offered an improved version of the Rubis class in use by the French Navy called Amethyste. British and French companies both looked for a Canadian base from which to launch their rival bids. DCN combined with the French national nuclear agency Tecnatome in order to bid for the submarine contract, forming SNA Canada Inc. Vickers has close ties to Canadian companies, notably MIL, which is a Canadian offshoot of the Vickers group.[46] The British company Plessey bought the Canadian electronics company Leigh Instruments in 1988 in order to bid on a variety of Canadian naval programmes including the sonar system for the SSN programme.[47]

The extent of the international industrial linkages involved in the submarine programme was not unique to CASAP but has also been a feature of other programmes. Each of the City class ships will be equipped with at least one helicopter of foreign origin. This seemed likely to be an anti-submarine version of the Anglo-Italian EH-101 aircraft, to be bought under the New Shipborne Aircraft (NSA) programme authorized in 1985. In 1987, EH Industries (an Anglo-Italian joint venture involving Agusta of Italy and Westland of the UK) won a contract to carry out the project definition for the new Canadian helicopter. Under this programme the aircraft was to have been built in Canada by a consortium of Canadian companies under the name of EH Industries Canada Ltd. The principal members of the consortium would be Bell Helicopters of Canada, Paramax Electronics and the Canadian Marconi company. It was intended that some of the avionics for the aircraft would be developed in Canada.[48] The feasibility study for the programme was commissioned by the Government in April 1988, when it was already clear that a successful bid would have to include a substantial Canadian input (possibly as high as 78 per cent of the value of the contract).[49] In 1989 this programme too was postponed indefinately by the DND.[50]

The naval import policy of the Canadian Government suggests that there has been an increased effort to use the ability to initiate large public sector programmes to support an industrial strategy. Whereas within market economies government intervention in civilian industrial planning is resisted by the business community, in the area of defence production there is no argument that this kind of planning is clearly a government responsibility. However, while the approach has been of potential benefit in terms of employment, in terms of improving the technology base there seem to be limits to what defence contracting can achieve. In the naval arena the companies primarily concerned with naval high technology—those associated with the development of sonar, radar, electronics and communications equipment—are by and large Canadian subsidiaries of US or European companies and are employing technologies developed elsewhere. Canada is the single largest market for US electronics exports while Canadian exports to the United States have also grown in the 1980s. Important Canadian contractors in this area of sub-systems include subsidiaries of Litton, Rockwell, Westinghouse of the United States (as well as Marconi and Plessey of the UK), and consequently a great deal of this trade takes place within the national subsidiaries of an international corporation.

Notes and references

1   *Challenge and Commitment, A Defence Policy for Canada* (Canadian Government Publishing Centre: Ottawa, 1987). The previous White Paper, *Defence in the 70s: White Paper on Defence*, was published in 1971.

2   Halstead, J., 'Canada's role in the defence of North America', *NATO's 16 Nations*, May 1985, p. 38.

3   See, in particular, *Canada's Maritime Defence*, the Report of the Sub-Committee on National Defence of the Standing Senate on Foreign Affairs, issued by the Minister of Supply and Services, Ottawa, Mar. 1983.

4   Dowd, E., 'Canada's navy is unable to defend its own coastline', *Daily Telegraph*, 21 June 1983; 'Reduced ASW ability of Canada in Ocean Safari', *Jane's Defence Weekly*, 19 Sep. 1987, p. 567.

5   *Defence 86*, Ministry of Supply and Services, Ottawa 1987, pp. xi–xii.

6   *Challenge and Commitment, A Defence Policy for Canada* (note 1), p. 49–50.

7   *Challenge and Commitment, A Defence Policy for Canada* (note 1).

8   Price, T. M., 'Canada's nuclear fleet: who will pay?', *Proceedings of the US Naval Institute*, Sep. 1987, p. 105.

9   Best, J., 'Canada to boost its defence spending', *The Times*, 12 Feb. 1988, p. 8; 'DND gets 2.7 per cent real spending increase', *Canada Defence Update*, Mar. 1988, p. 3.

10   The most detailed overview of Canadian procurement is Haglund, D. (ed.), *Canada's Defence Industrial Base: the Political Economy of Preparedness and Procurement* (Robert Frye and Co.: Kingston, 1988).

11   Winsor, H. and Harrison, M.,'High costs force Canada to cancel submarine order', *The Independent*, 28 Apr. 1989; Milner, M. and Sanger, C.,'Canadian cuts sink hopes of £3.5 bn in sub sales', *Guardian Weekly*, Week ending 7 May 1989, p. 3.

12   'Sub deal death knell?', *AAS-Milavnews*, Feb. 1989, p. 5; 'Canadian reshuffle may delay SSN', *Jane's Defence Weekly*, 11 Feb. 1989, p. 203.

13   Hobson, S.,'Mulroney ends doubt over SSNs for Canada', *Jane's Defence Weekly*, 18 Mar. 1989, p. 430.

14   *Canada's Defence News Bulletin*, 31 May 1989, p. 2.

15   'Hardware behind the paper', *NATO's 16 Nations,* special issue, Jan. 1988, interview with Eldon J. Healey, Canadian National Armament Director, pp. 94–95.

16   *AAS Airletter*, Feb. 1988, p. 44 .

17   Quoted in 'Nuclear submarine plan is most controversial in Canadian White Paper', *Defense & Foreign Affairs Daily*, 19 June 1987.

18   'Canada to boost coastal defence', *Jane's Defence Weekly*, 6 Aug. 1988, p. 206.

19   'TRUMP: Tribal Class update and modernization programme', *NATO's 16 Nations*, special issue, 1988, pp. 114–15.

20   Hewish, M., 'Trends in shipborne radar', *International Defense Review*, June 1988, p. 671.

21   Hewish, M., 'Canada's procurement programme', *International Defense Review*, Sep. 1987, p. 1230.

22   'Canada to build new frigates', *Financial Times,* 5 Jan. 1978.

23   'Shadwick, M., 'The Canadian Navy: Recovering from "rust-out"', *Naval Forces*, 1988, Naval Balance edition, pp. 66–67.

24   Shadwick (note 23); Friedman, N., 'Western European and NATO navies', *Proceedings of the US Naval Institute*, Mar. 1988, pp. 40–41.

25   *Defense News*, 30 May 1988; *Jane's Fighting Ships 1988–89* (Jane's: Coulsden, 1988), p. 74.

26   Couhat, J. and Baker, A. D. (eds), *Combat Fleets of the World 1986–87* (Naval Institute Press: Annapolis, Md., 1986), p. 44.

27   Price (note 8), p. 103–105.

28   Hewish, M., 'Canada's procurement programme', *International Defense Review*, Sep. 1987, p. 1230; 'Canadian SSN choice: more delays possible', *Jane's Defence Weekly*, 6 Aug. 1988, p. 196.

29   'Opposition to Canada's Defence White Paper', *Defence*, Sep. 1987, p. 506; Best, J., 'Canada to boost its defence spending', *The Times*, 12 Feb. 1988, p. 8.

30   'SSNs best option for Canada, says navy chief', *Jane's Defence Weekly*, 20 Feb. 1988, pp. 288–89. The case for a Canadian SSN fleet is argued in Tracy, N., 'Matching Canada's Navy to its foreign policy and domestic requirements', *International Journal*, summer 1983, pp. 459–76.

31   Watkins, Admiral J. D., testimony to the US Senate Subcommittee on Seapower and Force Projection, 14 Mar. 1984, reproduced in Tracy, N., 'Why does Canada want nuclear submarines?', *International Journal*, summer 1988, p. 514.

32   Sanger, C., 'Arctic deal ignores US submarines', *The Guardian*, 12 Jan. 1988. For a discussion of the issue of Canadian sovereignty in the Arctic, see Pullen,

T. C., 'What price Canadian sovereignty?', *Proceedings of the US Naval Institute*, Sep. 1987, pp. 66–73.

33   Polmar, N. D., 'Sailing under the ice', *Proceedings of the US Naval Institute*, June 1984, pp. 121–23.

34   Pullen (note 32), p. 72.

35   Reproduced in 'Nuclear submarine plan is most controversial in Canadian White Paper', *Defence & Foreign Affairs Daily*, 19 June 1987, p. 2.

36   'US Navy fears may snag Canadian nuclear sub buy', *Defense News*, 23 Nov. 1987; Ashford, N., 'US technology may bar Britain from Canada sub order', *The Independent*, 12 Oct. 1987, p. 13.

37   'France's Amethyste and the Canadian SSN procurement programme', *International Defense Review*, July 1988, pp. 741–42.

38   Brogan, P., 'Reagan allows Ottawa to buy British submarines', *The Independent*, 28 Apr. 1988.

39   Krass, A. S., Boskma, P., Boelie, E. and Smit, W. A., SIPRI, *Uranium Enrichment and Nuclear Weapon Proliferation* (Taylor & Francis: London, 1982), p. 238.

40   'Talks do not ease Canadian sub sticking points', *Defense News*, 27 June 1988, p. 4; 'Nuclear transfer agreement struck but final sub choice delayed', *Navy News & Undersea Technology*, 25 July 1988, p. 1.

41   Sassheen, R. S.,'Small but deadly submarine', *Asian Defence Journal*, Feb. 1989, pp. 70–71; 'AMPS: A miniature nuclear powerplant', *International Defence Review*, May 1988, p. 532.

42   Sassheen, R. S., 'Small but deadly submarine', *Asian Defence Journal*, Feb. 1989, pp. 70–71.

43   'Data sparse on SSN technology transfer', *Canadian Defence Update*, Mar. 1988, p. 1; 'French industry moves to fill Canadian sub needs', *Defense News*, 23 May 1988, p. 12.

44   *Canada's Maritime Defence*, the Report of the Sub-Committee on National Defence of the Standing Senate on Foreign Affairs, issued by the Minister of Supply and Services, Ottawa, Mar. 1983, p. 67.

45   'Canadian companies submarine alliance', *Jane's Defence Weekly*, 14 May 1988, p. 953; 'Potential SSN contractors consolidate', *Canadian Defence Update*, May 1988, p. 3.

46   Harrison, M., 'VSEL steps up battle for £4 bn Canadian submarine order', *The Independent*, 17 Dec. 1987; Preston, A., 'Countdown to CASAP', *Defence*, May 1988, p. 368; Harrison, M., 'VSEL hopeful of winning Canada order', *The Independent*, 17 June 1988.

47   'Plessey acquires Leigh Instruments in Canadian market', *Defense News*, 18 Apr. 1988, p. 30.

48   'EH industries to prepare copter replacement study', *Defense News*, 9 May 1988, p. 18.

49   'EH Industries in study for Canadian helicopter', *Jane's Defence Weekly*, 23 Apr. 1988, p. 769; 'EH Industries to prepare copter replacement study', *Defense News* (note 48).

50   'Canadians postpone helicopter decision', *Jane's Defence Weekly*, 28 Jan. 1989, p. 117.

8. India

I. Introduction

Perhaps more has been written about the political implications of military developments in India than about those of any other Third World country. Moreover, along with the development of the Indian nuclear programme, the growth of the Indian Navy has been seen as a development particularly worthy of attention. At issue has been the extent to which India's significant superiority in naval forces over its neighbours gives the Indian Government an effective tool with which to pursue regional hegemony in south Asia and throughout the Indian Ocean. A great deal of this large literature has been accusatory in tone and characterized by critical assessments based on assumptions about Indian intentions.

There is little doubt that successive generations of Indian leaders have shared an aspiration to play an expanded role in international politics. Nehru's belief that the international system is essentially multipolar and that the dominant role of any single global power is a transient phenomenon has been the intellectual underpinning for Indian foreign policy since 1947. From an Indian perspective, the emergence of a number of independent power centres in the world is inevitable. However, there are open questions about whether this group of independent power centres will include India and whether the relations between these countries will be competitive or co-operative. Indian leaders have also shared a sense of frustration that the major powers in general, and the United States in particular, have been indifferent or hostile to this Indian approach to international relations. In many instances Indian non-alignment has been defined as a thin veil for alliance with the Soviet Union. Moreover, Indian defence programmes have consistently been portrayed as contributing to the political instability in Asia, undermining the nuclear Non-Proliferation Treaty and hindering domestic economic development in the face of large-scale suffering among the majority of the Indian population. Conversely, Indian authors have often been defensive about Indian policy, seeking to depict Indian actions as reactive in the face of foreign threats and provocation or presenting India as an alternative model of development for Third World countries, distinct from either capitalism or socialism. Of the enormous volume of literature published on Indian military development, surprisingly little has begun from the premise that India is just another country pursuing national interests within the confines of what is objectively possible.

II. Indian naval programmes

India stands apart from the other countries under discussion because of the range and number of naval programmes currently under way. Unique among the national programmes discussed in this monograph, the Indian Government has committed itself to a major investment to increase the number of naval combatants heavily armed with anti-ship weapons. This programme is predicted to produce a navy of over 100 major combatants, about 50 of which will be larger than 3000 tonnes in displacement.[1]

There has consistently been a disparity between the level of naval development that the Indian Government has perceived to be needed and the level of development possible from indigenous resources. In spite of the existence of a major shipbuilding capability, the Indian Navy will rely heavily on imported ships as well as weapon systems for much of its planned increase in fire-power. Imports of major combatants are chiefly a consequence of major deals with the Soviet Union signed in 1980 and December 1982. These orders included three Kresta II class cruisers, though it seems that this deal may now have been cancelled, three Kashin class guided missile destroyers (in addition to three ordered previously). In 1984 or 1985, an undisclosed number of Kilo class submarines, now believed to be six, were ordered. This was probably the first export of this class agreed to by the Soviet Union, although a deal with Poland appears to have occurred virtually simultaneously. Kilo class submarines have been delivered to India by the Soviet Union at a rate of one per year since 1986.[2]

Although India has depended on foreign suppliers for the majority of major combatants, import programmes have proceeded parallel with new construction. Of the 13 naval programmes under way that will add major combatants to the fleet, 8 involve production in India.

Moreover, there is a considerable diversity among these programmes. Classes recently produced or currently in production include the Godavari class frigate; West German Type 1500 submarines; a new Indian-designed anti-submarine escort occasionally referred to as 'Project 15'; a large offshore patrol vessel of 1800 tonnes and three types of smaller patrol boat (one designed locally, the others in Japan and Singapore, respectively). The Godavari programme has been terminated after the completion of three (out of a planned six) ships, in favour of the new escort mentioned above.[3]

A continued investment in shipbuilding in India is planned. At present the Mazagon Docks and Garden Reach naval dockyards in Bombay and Calcutta, respectively, are primarily responsible for naval production, although the Vasco Da Gama shipyard in Goa is also an important defence contractor. Bombay has been the principal location for the construction of frigates and large offshore patrol vessels (OPVs). Production of a new class of escort with a displacement of around 5000 tonnes, probably intended for air defence, is due to begin in Bombay in 1989. In Calcutta the production of corvettes and large tank landing ships has been in progress, with some reports that the first of the latter have been launched.[4] The corvettes, displacing around 1200 tonnes, are built to an Indian design, although they incorporate a large number of foreign sub-systems, including the engines and weapon systems. Designated Type 25 or Khukri class in India, these are apparently armed with both surface-to-surface and anti-aircraft missile systems.[5] These vessels are planned to replace the Petya class corvettes currently approaching retirement from the Indian Navy. A new class of minehunter/sweeper is apparently to be ordered in Europe with two to be bought directly and six subsequently built at Goa Shipyards. Unconfirmed reports have suggested that this may be the Alkmaar class minehunter, the Dutch version of the tripartite minehunter produced under a NATO collaborative programme.[6]

Statements by the Indian Naval Chief of Staff further suggest that a replacement for the aircraft-carrier *INS Vikrant* would be built in India. However, there is much rumour and little information concerning this programme. During the evaluation which led to the purchase of the Hermes class aircraft-carrier (now named *INS Viraat*), Indian naval planners took a particularly close look at the Garibaldi class ships planned for the Italian Navy. However, it now seems that after Indian naval staff toured a series of prospective suppliers, including France, Spain, the United Kingdom and the United States, the preference of the Indian Navy was for French assistance.[7] In mid-1989, the Indian Government signed an agreement with the French government-owned shipbuilding consortium DCN (Direction Technique des Constructions Navales) to conduct feasibility studies into the design of a new aircraft-carrier.[8]

The programme will apparently be initiated by the early 1990s, and production of a vessel of around 30 000 tonnes displacement is planned for the Cochin shipyard, where there is already experience of the construction of 75 000-tonne bulk carriers and plans to build oil tankers of up to 86 000 tonnes for the Shipping Corporation of India.[9] While the quality of the facilities and the workforce has not been questioned, it is not clear that such an ambitious project will be possible in India. The initial cost estimate was US $400–470 billion To this will have to be added the fee for overseas production consultancy, initial plans to use an indigenous design having already been scrapped. Although the site is currently underutilized, Cochin has a full order book, and for all these reasons the timetable for any programme as complicated as the construction of an aircraft-carrier is uncertain. The *INS Vikrant* began a comprehensive modernization in the Mazagon Dockyard during 1987. Expected to take two years, this programme is intended to keep *Vikrant* in service through the 1990s and included replacing the catapult with a ski-jump and fitting a new radar suite.[10] Consequently, it seems probable that discussions of a new carrier are intended to lead to the acquisition of a replacement for *Vikrant* in the late 1990s or beyond. If this assessment is correct, it suggests that the Indian Navy will retain two aircraft-carriers, one based at the Western Fleet headquarters in Bombay, the other at the Eastern Fleet headquarters in Vishakapatnam.

The growth of the Indian submarine fleet includes the addition of conventional submarines of differing tonnages from the Soviet Union and FR Germany as well as the leasing of a nuclear-powered submarine from the USSR. This last programme includes the addition through a leasing arrangement of at least one nuclear-powered attack submarine from the Soviet Union, which may over the very long term preface the purchase of up to five more. Finally agreed during Rajiv Gandhi's visit to Moscow in 1985, this has made India the first country outside of the nuclear weapon powers to operate such vessels. Indian submariners have been receiving training related to nuclear-powered submarine operations from the Soviet Navy for two years.[11] Indian submarine programmes are discussed on page 144.

In the realm of maritime aircraft, India is one of two developing countries (the other is Argentina) that has developed significant shore-based naval air forces and shipborne fixed-wing aircraft. However, of the programmes in the two countries, those under way in India are considerably more extensive. The 1988 Indian annual defence review contains plans for an expanded naval aviation arm which would include 193 aircraft of all types, 80 of them ship-based.

III. Indian naval force planning and development

The political context in which these ambitious naval plans have been presented by the Government of India has not emphasized the possibility of any major confrontation with superpower or other major navies.

In a December 1988 interview, the Defence Secretary, the senior civil servant in the Indian Ministry of Defence, described the growth of the Indian Navy as follows:

All I can say is that the Indian Navy has improved tremendously in the past 10 to 15 years. But you should consider that we have 6 000 miles of coastline and an EEZ which is equal to almost two-thirds of the land mass of the country . . . I don't think that the extent of our naval modernization— number of ships, naval aircraft, ammunition aboard ships; any way you wish to look at it—gives us a very large navy compared with the task which the large coastline, the EEZ and the island possessions requires of us. Something like 60 to 70 per cent of the oil which we produce indigenously comes from offshore wells, these have to be protected also . . . So if you look at the enormous task which the Indian Navy has to undertake, and also at the turbulence of the Indian Ocean region and the growth of other navies in the region, then the growth of the Indian Navy is not a cause for alarm.[12]

The growth of the Indian Navy is not explained above in terms of a need to combat the forces of major powers deployed in the region. Equally, US perspectives on the Indian Ocean are not focused on operations against India. Rear Admiral Studeman, Director of US Naval Intelligence, in an intelligence briefing to Congress, noted that 'India is an unlikely antagonist of the US, but the proliferation of Soviet weapons in a region where we might meet the Soviets at least complicates the rules of engagement'.[13]

The potential for conflict between Indian and US naval forces would therefore lie in the context of a US–Soviet crisis, where US Navy rules of engagement would permit the engagement of any identified and potentially hostile vessels, rather than as a consequence of any bilateral Indo–US issue. There exists an important contrast between these official positions and a great deal of what is written in the media both inside India and about India. Much of this writing makes reference to Indian concerns about the possible threats to India arising out of US–Pakistani defence co-operation after the creation of the US Central Command and the Soviet invasion of Afghanistan.[14] While the Indian Government has certainly been critical of US assistance to Pakistan, the idea of a US–Pakistan 'axis' aimed at India is not a rationale for Indian equipment programmes. These programmes can be explained more easily in terms of India's perception of itself as a significant Asian power. This perception was already manifest in India at the time of independence.

The mechanics of procurement planning

Naval programmes as they have unfolded since the defence review conducted by Prime Minister Indira Gandhi after her return to power in early 1980 illustrate the nature of procurement choices faced by Indian decision-makers. Indian procurement has historically been the outcome of one of three choices: buy from Western Europe; buy from the Soviet Union; or produce weapons in India. The naval programmes currently under way in India illustrate the continuing importance of all three of these options.

The victory of Mrs Gandhi's Congress Party in the December 1979 Indian general election prompted a decision to review all procurement programmes initiated by the Janata Government.[15] In spite of predictions that Janata would give Indian foreign policy a more Western orientation, ties with the Soviet Union were not weakened between 1977 and 1979. Nevertheless, out of office Mrs Gandhi had been critical of decisions to buy military equipment in Western Europe. This equipment was not only expensive in its unit cost (especially compared to Soviet systems that could be paid for in rupees) but also disruptive when introduced into a force structure increasingly oriented during the 1970s towards the operation of Soviet equipment.

Long-term naval plans were accorded lower priority than Army or Air Force programmes in India until the mid-1960s and the publication by Indonesia of claims to Indian archipelagos close to South-East Asia (the Andaman and Nicobar Islands). This territorial claim was supported by the growth of what seemed at the time like a powerful Indonesian naval force based first on Soviet and subsequently on Chinese imports. Indian sensitivity to potential threats to remote and undefended territory had been heightened generally by defeat in the Himalayan war against China. It has subsequently been further enhanced by successive instances of 'island grabbing' both in South-East Asia (in the Paracel and Spratly archipelagos) and events such as the Falklands/Malvinas War.

The expansion of Indian maritime interests continued with the discovery of the Bombay High offshore oilfield. The discovery of offshore economic assets coincided with the Law of the Sea process, which opened the possibility further to increase economic returns from the sea.

The British decision to review and subsequently cut back its military commitments east of Suez also accelerated Indian interest in building more effective naval forces. The growth of regional military forces in the Persian Gulf region, including those of Iran and Saudi Arabia, as well as the expanding US and Soviet naval presence, reinforced the case for an increase in Indian submarine forces.

The long-term goal of Indian naval planning has been the construction of three roughly balanced fleets to cover the western, eastern and southern seaboards of the Hindustan peninsula, supported by a widened capacity for shipbuilding, repair and maintenance. This plan itself evolved during the 1970s to include the development of a large maritime air arm and the upgraded importance of the Indian Coast Guard in security planning. The growing involvement of maritime air forces and the Indian Coast Guard in Indian policy is discussed in a special section below.

Within this long-term evolution, the procurement of equipment was secondary to the creation of a shore-based infrastructure of bases and shipyards able to support a Navy of the size India has planned. Deliveries of weapon systems to India during the 1980s represent the fleshing out of an infrastructure constructed over a much longer period. The network has been based on naval facilities developed for the Royal Navy during the period of British rule, but has also involved the creation of entirely new facilities and, in particular, naval air bases.[16] The Indian Navy is now divided into three commands— Western, Eastern and Southern Commands—based at Bombay, Vishakapatnam and Cochin respectively. The Western Command will eventually move its headquarters to Karwar, midway between

Bombay and Cochin, once facilities under construction since 1986 are completed. These facilities will include a major new naval air station and a new shipyard. In 1972, facilities in the Andaman and Nicobar Islands in the eastern Indian Ocean, particulary the facilities at Port Blair, were selected for major upgrading. A long runway was under construction in 1986 on Great Nicobar Island, leading to complaints from the Indonesian Government.[17]

During this relatively long history, the overall naval equipment programme has gathered a considerable bureaucratic and political momentum behind it. In India, defence decision-making is ultimately the responsibility of civilians, either politicians or civil servants in the Indian Ministry of Defence, where there are no uniformed personnel. While the Defence Committee of the Cabinet regularly calls on the Chiefs of Staff of all three of the armed forces to give evidence and as advisers, the service Chiefs are excluded from decision-making. Moreover, the structure of Indian defence organization is such that there is no institutional framework for joint or integrated planning by the three services. There is no Joint Chiefs of Staff Committee, and consequently no Chief of Defence Staff, while there are many occasions on which the senior officers can meet one another, they are not compelled to do so and the level of co-operation and consultation between the three services is a function of the personal relationships between the individuals involved.[18]

Naval equipment programmes under way in India highlight the difficulty of dating the initiation of Indian equipment programmes precisely. Orders seem to have been the product of three 'bursts' of ordering in 1980, in 1982 and in 1985. Every year between 1980 and 1985 saw large military delegations from India visit the Soviet Union or vice versa. 1982, in particular, was a year of regular and important contact between senior political and military officials from India and the Soviet Union. In March, Marshal Dmitri Ustinov visited Delhi together with the Chiefs of the Soviet Navy (Admiral Sergei Gorshkov) and Air Force and the Deputy Chief of the Army. In October a large Indian delegation, led by the then Naval Chief of Staff Oscar Dawson, visited Moscow and later in the year the signing of the arms transfer package in December was preceded by a visit to Moscow by Mrs Gandhi—her first visit to the Soviet Union after returning to power in 1980.[19]

The planning of these programmes actually began before her return to power. Many of the arms programmes currently unfolding in India are the outcome of deals finalized in the turbulent five-year period 1978–82, when officials discerned a disturbing pattern emerging in regional politics in which serious domestic instability was followed by intervention. The 1978 coup in Afghanistan and the revolution in Iran were followed by the Soviet invasion of Afghanistan and the outbreak of the Iraq–Iran War, respectively, as well as the creation of the US Central Command and the changing nature of US relations with China. By 1980 there was already a perception that these regional developments might stimulate a more intimate Sino–US–Pakistan relationship, and one possible manifestation was that for the first time both the USA and China would support Pakistani arms programmes at the same time. Moreover, given that the Janata Government in India, much to its chagrin, was not even informed let alone consulted about the entry of Soviet forces into Afghanistan, there was concern that this might occur at a time of change in the Indo-Soviet relationship.

In this climate, one might have expected a greater emphasis on the development of land and air forces, where the greatest Pakistani investment was expected. In fact, the Navy has been very successful in competing for resources. The long-term development of naval plans was not disrupted after 1979–80 to meet the needs of the Army or Air Force. Rather, the defence sector as a whole was accorded a greater share of national resources. By 1987–88, contractual obligations undertaken during the period 1980–85 had insulated the Indian Navy from cuts in spite of some difficult choices between competing programmes faced by the Government. In the reduction of defence expenditure in 1988, undertaken after successive failures of the monsoon led to serious drought, Indian Air Force (IAF) programmes have been delayed (notably decisions on whether to proceed with licensed production of the MiG-29 and concerning a new jet trainer aircraft). Army and Navy programmes, including the Fleet Air Arm, seem to have remained intact.[20]

India does not and is not likely to face any threat of seaborne invasion. By contrast India would be extremely vulnerable to any disruption of seaborne trade. Far from self-sufficient in any category of goods, Indian dependence on the sea for its economic growth is growing rather than shrinking. While it has not been widely acknowledged in the literature, this fact has been reflected in naval development.[21]

The composition of Indian programmes, including the ship-based maritime air component, has always stressed anti-submarine warfare rather than other missions. The aircraft-carriers currently to be operated by India will offer a chance to deploy up to 40 helicopters in the vicinity of South-East Asian 'choke points' through which Chinese submarines would have to pass in order to reach the Indian Ocean. Equally, these aircraft could form part of a merchant-shipping convoy escort. Bréguet Alize aircraft have been replaced by Sea King helicopters on the *INS Vikrant*, and the *Viraat* will have the same aircraft complement.[22] Moreover, the acquisition of helicopters equipped for ASW has been a feature of other Indian naval programmes. The Godavari class frigates essentially modified the ex-Royal Navy Leander class frigate design to allow an additional helicopter to be embarked. Asked about the role of the carriers, the Indian Chief of Naval Staff said that the aircraft-carriers were intended for tactical battles at sea and were not intended to project any strategic motives.[23] Indian Defense Minister P. V. Narasimha Rao was in Moscow in mid-April 1985, where he stressed the need for 'layered defence' in which different ASW platforms would be able to operate from coastal waters far out into the Indian Ocean.[24]

The growing importance of maritime aircraft

India's Navy does not have complete responsibility for maritime air forces; the Air Force and the Coast Guard operate some of the requisite systems. However, the nature of the operational division of labour is unclear. Among the formal roles of the Indian Air Force is the tactical air support of the Indian Navy, and the Air Force has five coastal air bases under the control of the Air Force Southern Command, whose headquarters is itself on the coast at Trivandrum on the southern tip of the Indian peninsula. Rather confusingly, although the Southern Command of the IAF has formal responsibility for operations in and around the island territories of the Andaman and Nicobar Islands, the only air bases on these islands are under the control of the Navy. The IAF has been receiving Jaguar fighter-bombers since 1979, initially delivered directly from the UK and, since 1982, from Indian licensed production. The latest production is to be armed with the Sea Eagle anti-ship missile, and is formally assigned to the maritime strike role. As noted earlier, there is no institutional framework for joint or integrated planning by the three services. However, the implication of the equipment programmes outlined is clearly that in reality the Indian armed forces (including the Coast Guard) have increasingly come to rely on integrated operations.

The Indian Navy itself has considerable and growing air forces. As noted above, the aircraft-carriers are or will be equipped with a full complement of fixed-wing aircraft and helicopters when at sea (for most of the time these aircraft are stationed at naval air bases in Goa), but recent Indian imports have focused on shore-based reconnaissance aircraft and helicopters equipped for a variety of missions. Shore-based maritime reconnaissance responsibilities were transferred from the IAF to the Navy in 1975.[25] The Navy was initially supplied with old, second-hand Air Force aircraft and four Soviet Il-38 maritime reconnaissance planes. During 1988, India has taken delivery of more advanced Soviet aircraft—5 Tu-142s, used by the Soviet Union as a communications relay and for ASW missions as well as reconnaissance. It is probable, but not certain, that Indian Tu-142s will be equipped with ASW torpedoes. The Indian Navy is also to receive some of the Dornier 228 maritime reconnaissance aircraft produced as a result of a deal signed in 1983 with the Federal Republic of Germany, involving the direct sale and subsequent licensed production of this aircraft in India.[26] Aircraft will be supplied to India's domestic air cargo carrier *Vayudoot* as well as the Indian Coast Guard and Navy, with naval versions being equipped with French radars and electronics as well as French AS-15 air-to-surface missiles.[27] Most of the aircraft operated by the Indian Navy are helicopters. These have been bought in a variety of versions intended for ASW, anti-shipping missions and airborne early warning. Similar to the policy pursued in the purchase of Dornier 228, the import of helicopters by the Navy was co-ordinated with other maritime users, such as the Oil and Natural Gas Commission. This increased the bargaining power of Indian negotiators in discussions with British and French suppliers. The British company Westland needed to win the order particularly badly, which further assisted India in winning advantageous financial terms from the UK including long-term low interest credit. The joint purchase also had the advantage of ensuring a degree of commonality in maintenance and resupply between the Indian Navy and facilities on offshore oil rigs.[28]

The Coast Guard is the third component of Indian maritime air forces. In 1975 a Committee was established in India to assess a study of Indian maritime air forces based on a case study of the US experience and organization. The allocation of responsibilities for reconnaissance missions to the Navy noted above was one consequence of the report, the other was the recommendation for the establishment of a Coast Guard. The establishment of a Coast Guard was closely related to the passage of the Territorial Waters, Continental Shelf, Exclusive Economic Zone and other Maritime Zones Act in 1976. The Act, which came into force on 1 January 1977, established a joint EEZ with Sri Lanka, of which 2 million km^2 was under Indian jurisdiction. Initially the Coast Guard was equipped with aircraft leased from civil airlines, but the purchase of maritime surveillance aircraft was acknowledged as a priority. As noted above, the Coast Guard was subsequently included in the programme for the acquisition of Dornier 228 aircraft, of which the service will ultimately receive 36. The Coast Guard was, under certain conditions, to come under the operational control of the Navy, but its primary responsibilities were surveillance of India's territorial waters and just beyond, and constabulary functions in that sea space. Beyond this coastal strip the Navy has responsibility for policing the EEZ.

The surveillance and monitoring function developing within the Coast Guard forces highlights one area where a civil or paramilitary programme will make a significant contribution to India's defence policy. Another area is the development of the Indian space programme.

India has invested significantly in the space programme—the declared space budget is currently US $176–200 million per year—with the intention of developing both satellites and launch vehicles.[28] Since 1975 India has been launching indigenously designed satellites for both communications and for ocean and land surveys. These satellites have been carried by Soviet, US and European launchers as well as Indian launchers. In 1980 India conducted its first successful launch of an indigenous rocket—the SLV-3—into a low orbit. Attempts to launch the successor to SLV-3—ASLV—failed twice in March 1987 and July 1988, but development of both ASLV and a larger launch vehicle—PSLV—continues.

Indian submarine programmes

The delivery of *INS Chakra*, a Soviet-produced Charlie class nuclear-powered submarine, to India in January 1988 was one development that has focused increased attention on Indian naval programmes. While India has operated a submarine fleet since the late 1960s, this development has been interpreted as a major change in Indian capabilities and evidence of India's intention to develop further its naval superiority among the countries around the Indian Ocean.

The origins of the SSN programme actually reach back almost 20 years. The Indian Navy received eight Soviet Foxtrot class diesel submarines after 1968, and discussion of their ultimate replacement was under way from the early 1970s.[29] This discussion essentially produced two conclusions: that in the long term India should aim to build submarines in the country, but that in the immediate future this would not be possible. Consequently there would have to be an overseas purchase to meet the needs of the Indian Navy in the interim. The discussion included the possibility that India would at some point want to build a nuclear-propelled submarine, euphemistically called the 'advanced technology vessel'. In December 1983, answering questions in the Indian Parliament, Defence Minister R. Venkataraman said: 'I have already said that we keep our options open in this matter; if necessary we will go in for it. But then a nuclear-powered submarine is different from the nuclear submarine with nuclear warheads. I have already said that we are not going to use atomic energy for anything but peaceful purposes. Therefore, we will use it for the power . . . It will be only propulsion'.[30]

It was clear by the late 1970s that the shipbuilding expertise, the shore-based support facilities and the manpower needed to build and operate nuclear submarines were lacking in India and could not be acquired quickly. At this time, the Soviet Union was apparently reluctant to transfer either nuclear-powered submarines themselves or the technology required for their construction in India. The Indo-Soviet arms deals concluded in 1980 and 1982 both included a naval component, including large surface combatants such as the Kashin and Kresta class ships and smaller surface vessels, patrol corvettes and mine countermeasure ships. However, the only submarines offered seem to have been refurbished Foxtrot class.[31] As a result, India began evaluating possible conventionally powered submarines to replace the Foxtrot submarines in service. At this point it was already decided that at least some units of the design which was chosen would be built in India, and that the ultimate objective of producing nuclear-powered submarines would not be sacrificed.[32] From the mid-1970s, a number of submarine designs were under consideration from Western Europe and the Soviet Union.[33] European countries involved were France, the FRG, Italy, the Netherlands and Sweden, with the FRG and Sweden the clearly favoured options by 1980.[34] Indian officials were looking for a design which could offer a chance to learn the production and operating skills relevant to operating nuclear-powered submarines. As noted in the discussion of Argentina, the Type 209 design offered by the West German company HDW met some of these criteria. In 1981 HDW won the order based on a 'stretched' and heavier version of the Type 209 weighing 1500 tonnes (and subsequently designated the Type 1500). However, this submarine is still some 300 tonnes lighter than that being built for the Argentine Navy.[35] FR Germany also gained an advantage in negotiations by offering as a package a new generation of torpedoes supplied by the West German company AEG.[36] The initial order covered the sale of two submarines to be built in Kiel and included an option to produce up to four subsequently in India. The signature of the contract was held up, as officials in the FRG were unhappy about a clause in the contract insisted upon by India which would guarantee deliveries of spare parts in wartime.[37] However, the option on the production of the submarines at Mazagon Docks in Bombay was exercised in December 1981.[38] Construction began in early 1982 and the West German-built vessels were delivered in 1986–87. Production of the submarines in India has run into problems, finally getting under way in 1984, and the first of these (originally expected in 1988), may now not be delivered to the Navy before 1991.[39]

In early 1984 there were reports of discussions with the Soviet Union on the supply of more advanced, possibly nuclear-powered, vessels and the training of Indian crews in the Soviet Union.[40] By late 1984, the Soviet Union was apparently prepared to offer India submarines of more modern design in considerable numbers. Vice Admiral Tahiliani, then Vice Chief of Naval Staff, took a leading role in talks in Moscow in September 1984 after which official sources stated that the defence relationship had taken on 'a new dimension'. This has subsequently been interpreted to have meant that the Soviet Union agreed not only to supply more modern types of conventional submarine, but also to allow India access to nuclear-powered submarines.[41] The formal agreement to lease a nuclear-powered submarine from the Soviet Union appears to have been signed in 1985.

The plan to develop an SSN force in India has not run smoothly, and highlights the enormous technological barriers for a developing country in operating this kind of system. To begin with, the shore-based facilities needed for nuclear submarines are significantly more complex than those for conventional submarines because of the need for reactor maintenance. There is currently no harbour facility in India capable of handling radioactive materials, and the submarine reactor is shut down when the submarine is in port. India has built a Soviet-designed facility called the Special Safety Service at Vishakapatnam, designed to monitor the health of people working on the *INS Chakra* and detect any radiation leak eminating from the submarine.[42]

In 1988 it appeared that the Indian Navy reached a plateau in terms of new orders for naval vessels. Existing contracts will lead to further deliveries over the next few years. However, the current focus seems to be on integrating the new equipment into the Navy. This has been explicitly elaborated by the Chief of Naval Staff, Admiral Nadkarni, who has stressed in interviews that while such long-term goals as the creation of a third fleet remain on the agenda, there is no prospect that these will be pursued in the immediate future.[43]

IV. Indian naval programmes and regional power

After the delivery of the second aircraft-carrier in 1987 an increasing amount of foreign attention has been focused on the growth of the Indian Navy. This has often been linked by analysts to two different developments in Indian foreign policy. The first is the emergence of a belief in India that New Delhi has a legitimate role in the domestic political development of countries around the Indian Ocean. The second is the claim that non-regional powers, and in particular the superpowers, do not have legitimate reasons for deploying military forces in the Indian Ocean. This has been called an Indian version of the Monroe Doctrine, or the Rajiv Doctrine.

The acquisition of a second aircraft-carrier and the SSN programme in particular have tended to shift the balance of analyses towards the role of the Indian Navy as a regional power-projection force.[44] Recent political developments have contributed to the picture of an increasingly assertive Indian foreign policy. In mid-1987, after 19 civilian Indian ships carrying supplies for Sri Lankan Tamils had been turned back by the Sri Lankan Navy, the Indian Air Force dropped food and supplies over Sri Lanka's Northern Province.[45] On 3 November 1988, Indian naval forces responded to a request for assistance from the Government of the Maldives for assistance against a coup attempt supported by Tamil mercenaries from Sri Lanka.[46] On 23 March 1989, on the expiry of the bilateral trade agreement between

India and Nepal, the Indian Government announced that trade between the two countries would be suspended until a new agreement had been negotiated.[47]

Reserving the right to intervene in neighbouring countries is clearly one strand of Indian policy. While both Indian and international opinion was generally supportive of Indian actions in the Maldives, Prime Minister Gandhi asked some Parliamentary critics: 'Are we to sit back and watch a democratically elected government of a friendly, neighbouring country being pulled down by alien forces?'.[48] However, this is not a new development. Apart from wars with Pakistan over possession of Kashmir, India also includes territories integrated by force. The Princely States of Junagadh and Hyderabad were incorporated in 1947 and 1948. In 1961 the former Portuguese colony of Goa was formally integrated into India and in 1975 the Himalayan Kingdom of Sikkim was annexed. In framing Indian foreign policy, successive governments have always been prepared to consider the use of force, and have intervened throughout the sub-continent where it has seemed that the use of armed force will be successful.

In spite of this history, Indian naval policy does not stem from any single cause, but responds to four broad pressures.[49] These are first, protection of the coastline and maritime approaches, including island territories; second, protection of natural resources in the Exclusive Economic Zone; third, protection of maritime trade and assets (including a merchant marine of over 700 vessels); and fourth, naval diplomacy/intervention or the promotion of regional political interests.

Moreover, the development of naval forces has not occurred in a vacuum but has been part of changes in India's bilateral and multilateral political relationships. On the one hand, in the 1980s the Indian Government has changed its broad economic policies in a manner which has allowed expanded relations with the United States and West European countries. On the other hand, the creation of the regional organization South Asian Association for Regional Co-operation (SAARC) has opened a multinational regional forum for discussions between governments for the first time in South Asia. These simultaneous developments in the economic, military and political spheres have created choices for the Indian Government about how it will respond to regional political developments. It is not so that India must respond militarily to regional developments, but it has that option. How frequently this capability will be used is currently impossible to predict.

Notes and references

1   Tellis, A. J., 'India's naval expansion: structure, dimensions and context', *Naval Forces*, May 1987, pp. 36–50.

2   'Third Soviet Kilo class sub for Indian Navy', *Times of India*, 25 Jan. 1988; 'Indian Navy's new Kilo submarine in close-up', *Jane's Defence Weekly*, 2 Apr. 1988, p. 639.

3   *AAS-Milavnews*, Mar. 1985, p. 22.

4   Jacobs, G., 'India's changing naval forces', *Navy International*, Feb. 1986, pp. 120, 124; 'Indian naval construction set for major expansion', *International Defense Review*, Mar. 1986, pp. 369–70; *Pacific Defence Reporter*, Aug. 1987, p. 36; 'Naval developments', *Defense & Foreign Affairs Daily*, 15 Dec. 1986, p. 2.

5   'Indian naval construction set for major expansion' (note 4); 'Challenges facing the Indian Navy', *Times of India*, 24 Dec. 1987, p. 3.

6   'Indian naval construction set for major expansion' (note 4); *Pacific Defence Reporter*, Aug. 1987, p. 36.

7   *International Defense Review*, Apr. 1985; Elliott, J., 'Swan Hunter seeks Indian Navy contract', *Financial Times*, 24 Apr. 1986; 'India to seek US aid in CV design', *Defence*, June 1988, p. 426; 'India to make aircraft carrier', *Hindustan Times*, 3 Dec. 1988.

8   'French design for aircraft carrier', *The Tribune* (Chandigarh), 13 July 1989.

9   *Times of India*, 10 Dec. 1987; 'India to make aircraft carrier', *Hindustan Times*, 3 Dec. 1988.

10 Sharma, L. K., 'Last modernization for INS Vikrant', *Times of India*, 30 Nov. 1987; Sen, M., 'Soaring beyond a floating township', *Times of India*, 28 Dec. 1987.

11 'Soviets agree to provide India with nuclear powered submarine', *Defense & Foreign Affairs Daily*, 7 July 1987, p. 1; Elliott, J., 'India may purchase Soviet N-sub', *Financial Times*, 15 Dec. 1987, p. 6; 'India and Brazil take first steps in acquiring SSNs', *Jane's Defence Weekly*, 19 Dec. 1987, p. 1399; 'Indian Navy enters N-age', *Times of India*, 7 Jan. 1988; 'New members for SSN club', *Jane's Defence Weekly*, 9 Jan. 1988, p. 11; 'Indian Navy has wider reach', *Times of India*, 12 Jan. 1988; Beaver, P., 'Indian SSN departs Vladivostok submarine base', *Jane's Defence Weekly*, 23 Jan. 1988, p. 116; 'Indian acquisition of nuclear powered submarines confirmed', *International Defense Review*, Feb. 1988, p. 108; 'India joins N-maritime club', *Times of India*, 4 Feb. 1988; 'Indian Navy greets first nuclear powered submarine', *Jane's Defence Weekly*, 13 Feb. 1988, p. 241.

12 Secretary of the Ministry of Defence of India, T. Seshan, interviewed in *Defense & Foreign Affairs*, Dec. 1988, pp. 18–22.

13 Prepared statement of Rear Admiral William O. Studeman, Director of Naval Intelligence US Navy, before the Seapower and Strategic and Critical Materials Subcommittee of the House Armed Services Committee on Intelligence Issues, 1 Mar. 1988, p. 27.

14 The differences in perception are typified by this exchange between an Indian and a US participant in a conference on the Indian Ocean:
    'Indian: you say the carrier battle group is for the Gulf, but we are concerned when Ambassador Hinton (formerly US Ambassador to Pakistan) leaves the door open for US involvement on Pakistan's side in an Indo-Pakistan war, as he did in his October 10 speech in Lahore.

'American: I would have said pretty unequivocally that the US would not, under almost any circumstances, get involved. I still think that is an extremely remote possibility', quoted in *India, the United States and the Indian Ocean*, Report of the Indo-American Task Force on the Indian Ocean (Carnegie Endowment for International Peace, Washington, DC, 1985), pp. 47–48.

15   The review was particularly aimed at the decision to buy the Jaguar strike aircraft: *AAS-Milavnews*, Apr. 1981; 'Indians ready to ditch low flying Jaguar', *The Guardian*, 12 Aug. 1980; 'India reviews agreement to buy UK Jet Fighters', *International Herald Tribune*, 21 July 1980; 'India seeks a review of British fighter plane pact', *New York Times*, 20 July 1980; 'Indian threat to scrap British planes order', *Daily Telegraph*, 18 Jan. 1980.

16   Portuguese facilities in Goa, until 1961 an enclave under Portuguese national control, have also been incorporated.

17   'New INAS bases', *Milavnews*, June 1986, p. 15.

18   The advantages and disadvantages of creating a Chief of Defence Staff are regularly argued in India. For a recent exchange, see the articles by Ashok Kapur in *The Statesman*, 25–26 Oct. 1988 and 26–27 Oct. 1988; and the responses by Lt-General S. K. Sinha in the same newspaper on 25 and 26 Nov. 1988.

19   *Daily Telegraph*, 10 Mar. 1982; *Daily Telegraph*, 2 Sep. 1982; *Daily Telegraph*, 26 Oct. 1982.

20   'Indian budget cut hits IAF programs', *Asian Aviation* (Singapore), Apr. 1988, p. 47.

21   One discussion of Indian attitudes that does exist is *India, the United States and the Indian Ocean*, Report of the Indo-American Task Force on the Indian Ocean (Carnegie Endowment for International Peace: Washington, DC, 1985).

22   'Indian retires Alize ASW from front-line service', *Jane's Defence Weekly*, 10 Oct. 1987, p. 797.

23   'INS Viraat arrives in Bombay', *Asian Defence Journal*, Sep. 1987, p. 127.

24   'Significant USSR–Indian deal discussed', *Jane's Defence Weekly*, 27 Apr. 1985, p. 704.

25   Singh, P., 'The Indian Navy', *Asian Defence Journal*, July 1987, p. 12.

26   'Dornier's cooperation efforts in India', *Times of India*, 14 Mar. 1988.

27   'Navy to go ahead with Dornier planes', *Hindustan Times*, 31 May 1987.

28   Elliott, J., 'Westland back in the running for Indian helicopter contract', *Financial Times*, 9 Sep. 1985; Elliott, J., 'India set to sign £220m order with British aircraft makers', *Financial Times*, 10 Oct. 1985; Elliott, J., 'India set to buy helicopters from UK and France', *Financial Times*, 17 Oct. 1985; Sharma, K. K., 'Britain wins Indian order for 21 helicopters', *Financial Times*, 27 Dec. 1985; 'India restates plan to buy Westland helicopters', *Aviation Week & Space Technology*, 6 Jan. 1986, p. 25.

28   The best overview of the Indian space programme is contained in *Jane's Spaceflight Directory 1988-89*, pp. 31–33 and pp. 442–44.

29   Manoj Joshi, 'The case of the Rs. 30 crore agent', *Frontline*, 18 Apr.–1 May 1987, pp. 11–12.

30   Venkataraman, V., *Parliamentary Debates, Rajhya Sabha*, Official Report, Part I, 6 Dec. 1983, pp. 11–12.

31   Jacobs, G., 'India's expanding naval forces', *Defence Today*, pp. 75–78.

32   'India is shopping again . . . this time for submarines', *Hong Kong Standard*, 6 Jan. 1979; 'Decision soon on submarines', *IDSA Strategic Digest*, Mar. 1979; 'No decision yet on sub project', *IDSA Strategic Digest*, Apr. 1979.

33 'India makes submarine decision', *Maritime Defence International*, Sep. 1980, p. 329.

34 Loudon, B., 'Indian sub order not for Russia', *Daily Telegraph*, 29 Apr. 1980. Details of the competition for the submarine order have emerged as a consequence of wider investigations into arms procurement in India sparked by allegations of corrupt practice surrounding the submarine contract and the contract with Bofors of Sweden for the supply of 155-mm howitzers to the Indian Army. See especially Manoj Joshi, 'The problem of the political agent', *Frontline*, 2–15 May 1987, pp. 15–16.

35 'India: German sub buy', *Defense & Foreign Affairs Daily*, 27 Mar. 1981, pp. 1–2; 'INS Shishumar commissioned', *Asian Defence Journal*, Dec. 1986, p. 74.

36 'Navy acquires 14th submarine', *Times of India*, 15 Mar. 1988; 'Choices facing the Indian Navy', an interview with Chief of Naval Staff Admiral Nadkarni, *Times of India*, 24 Dec. 1987, p. 3.

37 'India: German submarine snag', *Defense & Foreign Affairs Daily*, 11 May 1981. This clause in the contract has subsequently proved controversial.

38 Sharma, K. K., 'West Germany to build submarines for India', *Financial Times*, 2 Dec. 1981.

39 Manoj Joshi, 'Run silent, run deep', *Frontline*, 12–25 Dec. 1987; Singh, P., 'The Indian Navy', *Asian Defence Journal*, July 1987; 'India to manufacture W. German submarines', *Financial Times*, 17 Feb. 1984.

40 *AAS-Milavnews*, May 1984, p. 22.

41 'Defence and disarmament review', *IDSA Strategic Digest*, Nov. 1984, pp. 1342–43; 'India negotiates for new arms from East and West', *International Defense Intelligence*, 7 Jan. 1985, pp. 1–2.

42 'Lab to monitor radiation safety of nuclear-powered subs', *Patriot*, 19 Dec. 1988.

43 Mukherjee, D., 'Indian Navy's focus is on consolidation', *Times of India*, 24 Dec., 1987; 'Challenges facing the Indian Navy', *Times of India*, 24 Dec. 1987, p. 3.

44 See, for example, Tellis, A., 'Aircraft carriers and the Indian Navy: assessing the present, discerning the future', *Journal of Strategic Studies*, June 1987.

45 'Indian boats return as Lanka refuses entry', *Times of India*, 4 June 1987; Sanjiv Prakash, 'Sri Lanka: India's embarrassment', *Defense & Foreign Affairs Daily*, 19 June 1987, p. 3; Elliott, J. and Sharma, K. K., 'Indian aircraft drop supplies in Sri Lanka', *Financial Times*, 5 June 1987.

46 Bilveer, S., 'Operation Cactus: India's prompt action in Maldives', *Asian Defence Journal*, Feb. 1989, pp. 30–33.

47 Salamat Ali and Kedar Man Singh, 'Economic squeeze', *Far Eastern Economic Review*, 30 Mar. 1989, p. 26.

48 Quoted in Bilveer, S., 'Operation Cactus: India's prompt action in Maldives', *Asian Defence Journal*, Feb 1989, p. 32.

49 In fact, the plans to develop a balanced force structure in India independent of external actors in terms of both the supply and maintenance of equipment date back to a Defence Review conducted in 1949 in the wake of the first Indo-Pakistan war. In this long-term force structure plan the needs of the Navy were accorded the lowest priority in the prevailing threat assessment prepared by the Indian defence establishment.

Part III. Conclusions

9. Conclusions

The focus of this monograph is international procurement programmes—that is, those involving the transfer of either equipment or technology between countries—rather than purely domestic processes. There are unique features of naval procurement policies which respond to the specific political, geographic and historic environment in which those policies are framed. Despite the diversity of the countries surveyed, it is possible to make some valid observations about general patterns within the international naval procurement process that have emerged from the cases discussed in this book.

Naval forces are a status symbol, and in all the cases studied the special symbolic value attached by the respective governments to the development and maintainance of naval forces emerged as a rationale for investments in equipment. Naval forces are not regarded simply as supportive of a military strategy; they have been used to signal a desire to be a significant independent actor in international affairs. This factor applies regardless of the level of economic development and is equally true for Argentina and India as it is for any of the major naval powers.

While it is clearly a real and important influence, the precise role of this symbolism as a factor in decision-making on naval affairs is difficult to quantify. The cases of Argentina, Brazil and India illustrate the influence of a country's maritime history and past naval development on perceptions of naval priorities and future programmes. The limited nature of current Brazilian naval programmes and the fact that none have been initiated in the past five years appear to have had no impact at all either on the general self-perception in Brazil that over the medium and long term the country must continue to be a regional military power or on the self-perception of the Brazilian naval establishment, which continues to talk in terms of nuclear-powered submarines and the construction of carrier-based battle groups. In Argentina and Brazil, as much as in India, there has been a belief that large and balanced military forces are a prerequisite to establish the natural place of the country within the international system. This has exerted a sustained pressure on defence decision-making. All these countries have also been motivated by interlocking (and mutually supportive) perceptions of their growing importance as regional powers and efforts further to enhance their international status. In Argentina, Brazil and India the influence of this factor of self-perception has waned during periods in which other priorities have risen on the agenda. However, it has never disappeared totally and is always reasserted after some time.

I. Naval arms transfers and naval power

As a result of the naval arms trade there has been a horizontal spread of new naval capabilities. However, this process has been less rapid and less widespread than is generally claimed. The major importers are largely countries with a long naval tradition, and are countries located in key regions of Asia and Latin America, where the major powers have defined major foreign policy interests. It is the coming together of these two developments—weapon proliferation in areas of major power interest—that has been a cause of concern. Moreover, this concern has been further fuelled by the realization in major naval powers that domestic budgetary considerations are likely to constrain the size of their future naval programmes. Therefore, the transfer of naval equipment and technology may have contributed to alterations in the maritime environment by providing some developing countries with a degree of 'immunization' against coercion.

The studies conducted here do suggest that there has been a change in the utility of naval power, but also that the speed and extent of that change have been very limited. The changing utility of naval power is not a direct function of technological development. The problems faced by major powers in the projection of force are as much political and economic as military or technical. Certain kinds of naval intervention remain technically feasible, but they now carry greater political, economic and psychological risks. Moreover, it is worth noting that the deterrent function of major power navies as well as their capacity for intervention can have important implications for coastal states and the risk of conflicts of interest, if not actual use of force. Not least, regional naval capabilities have grown against a backdrop where the forces of the major powers (in particular those of the superpowers) are not only deployed world-wide but, on occasion, operate in deliberately close proximity to one another. These operations have increased during the 1980s in both their number and in the extent to which naval forces are prepared to challenge, even provoke one another. This is largely but not exclusively an issue for superpower navies. The primary responsibilities of most NATO navies lie in fulfilling alliance obligations, and therefore in acquiring weapon systems capable of operating effectively in the waters contiguous with Europe or the North Atlantic. However, some medium powers—notably Britain and France—require a naval force structure that does not totally foreclose the option of conducting operations outside the NATO context. The impact of this priority for 'out of area' operations on procurement policy has been limited—the operational deficiencies of the Royal Navy outlined by Parliamentary investigations conducted after the Falklands/Malvinas War were partly a function of the extent of the influence of alliance considerations on procurement policy in that the Royal Navy is now equipped primarily for anti-submarine operations and not for power projection.

A different but related issue concerns the opportunities that naval programmes offer specific countries, those whose geography places them close to or in control of important straits or 'choke points'. For these countries larger naval forces may create the opportunity to dictate the terms on which passages remain open to shipping. This issue revolves around whether use of the seas is a common good that cannot be denied or whether individual countries ought to be allowed to regulate maritime activities of non-national shipping. Freedom of navigation is not an issue of contention in most maritime countries, which are themselves dependent on the free passage of seaborne trade. Traditional regional naval powers have had the technical capacity to challenge the users of regional waters for at least the past 10 years. However, in these countries the implications of any form of confrontation have been held to be counter-productive. Any direct challenge to the US Navy would not only invite a military response that could not be dealt with easily, but it would also risk a range of important political and economic links to the United States and its allies. There is no evidence that coastal states, generally, are considering a challenge to the freedom of navigation, implying as it does a confrontation with the major naval powers. Potential clashes over freedom of navigation are restricted to countries whose geography puts them in control of 'choke points' or straits, such as Iran and Indonesia.

Neither of these countries is undertaking expansive naval programmes; and in the long term these countries are perhaps most dependent of all on keeping open the strategic waterways that they control, since closure would upset their own economies more than anyone else's. However, there is a question surrounding whether these countries can legitimately impose terms on which straits will remain open. It is largely attacks on merchant shipping in the Persian Gulf that have kept this issue at the centre of global attention. Events of 1987 and 1988 have demonstrated that, should they so wish, strategically located countries may threaten to close straits with a combination of sea mines, aircraft and coastal artillery (including coastal missile batteries) as well as with mobile naval vessels. The international mine-clearing operation and the escort of tankers through sections of the Persian Gulf have underlined that for the moment at least the international community has the technical capacity to oppose the imposition of regulation on straits passage successfully. The clarity of this message was reduced by the one-sided nature of the operations, which focused on Iran and excluded one of the principal offenders—Iraq—from sanction.

While the central confrontation between the major alliances has dictated that naval forces be constructed in the context of strategies of deterrence, the navies of major powers have never been confined to this role. In particular, naval forces have been a central element in policies aimed to assure access to the imported raw materials, in general, and oil, in particular, on which industrial economies are dependent. While oil production by new suppliers—notably Mexico, Norway and the United Kingdom—reduced this dependency by 1986, over the long term the place of the Persian Gulf littoral countries as suppliers will be reasserted. Whether any of these countries singly or in combination can keep Straits closed in the face of a response by the major naval powers remains an open question. However, the fact that there is any question at all is in itself evidence of a change in the global environment, where major powers used to exercise dominant military force. It may mean in turn that issues such as freedom of navigation will be seen less as naval problems than problems of regional policy in the Persian Gulf.

Changes in the naval equipment market have also occurred as a consequence of changes in the maritime law, in particular the creation of the 200-nautical-mile EEZ and the opportunity to exploit offshore resources. The creation of these zones does represent a major change in attitudes to the use of the sea, in particular in the Pacific and in the South Atlantic. Many isolated and sparsely populated islands that were previously regarded as of little economic value, perhaps even seen as burdensome to the government, have now been recognized as valuable pieces of real estate because they allow the demarcation and exploitation of a large EEZ.

Claims to these resources need to be reinforced by forces under national control, and compliance with the management scheme established by coastal states can only be ensured through effective surveillance and policing. Moreover, the creation of the EEZ places on coastal states an obligation to regulate and manage offshore economic activities in an orderly fashion, and countries who argue that their needs receive too little international consideration can demonstrate that they are able as well as willing to play a wider role in the international system by discharging the responsibilities implied by the LOS Convention effectively.

It is fair to conclude that while decision-making on naval arms transfer programmes reflects concerns about the political economy in both buyer and seller, ultimately the decisions about who to buy from and who to sell to are determined primarily by factors other than profit maximization, cost effectiveness or economic management.

II. The naval weapon acquisition process

It was appropriate to begin the conclusions with a political discussion because the basic justifications for naval equipment programmes lie in perceptions of national security interests. However, the extent to which government perceptions of purely national interest can be pursued is limited by the impact of three basic groups of factors; first, the shape and size of existing naval forces and the associated naval infrastructure; second, questions of the government budget and the national economy; and third, questions of alliance and alignment.

The degree of past investment in naval infrastructure is a powerful factor promoting continuity in the shape of naval forces. Keeping even a relatively small number of ships at sea involves a complex support system of bases and shipyards for replenishment and repair. The capital costs of developing this support network are considerable and the people needed to man it further contribute to increased defence expenditure. However, once the infrastructure to support naval forces has been developed, it acts as a factor which tends to preserve naval forces at a certain level. The fact that so much capital and so many jobs are tied up in sustaining a fleet reinforces the arguments for modernizing without disrupting the underlying structure. Even if this does not lead to replacement of naval vessels on a one-for-one basis, it may at least increase the incentives to seek vessels of a similar tonnage and characteristics. This pattern is not unique to naval force structures; it applies also in the areas of land and air forces. However, the scale of and complexity of 'big-ticket' equipment programmes intensifies its importance in the naval area.

The need for this extensive infrastructure means that the balance of naval forces is not something which can be changed over a short period. The Indian Navy, built around naval facilities constructed for the Royal Navy and operating, almost exclusively, vessels of British origin before 1968, took almost 20 years to learn how to cope fully with the different operational and maintenance requirements of Soviet vessels. The problems of integrating new equipment and the advantages of compatibility in maintenance and operation of equipment have made the major importers of naval systems reluctant to change suppliers, and relationships in the sphere of naval equipment supply tend to be long-term and durable.

The great majority of coastal states find that long-term force planning is difficult or impossible. Without a significant domestic defence industrial base it is necessary for these countries to import virtually all their requirements. In many cases, projecting future economic performance is difficult, which means that short-term availability, opportunism and reactive policies characterize the procurement process in developing countries the economic circumstances of which are too uncertain to allow any long-term planning. Among these countries there is a need for procurement strategies which require minimal capital expenditure. There are alternative options available to these countries, and the course most often pursued is reliance on second-hand equipment to fill their requirements. In some cases, such as that of Soviet transfers to Peru, this equipment has not been been accompanied by any formal political linkages. However, in the overwhelming majority of cases the extent to which relationships in the naval sphere have mirrored broader bilateral relationships in the area of security policy is striking.

Medium-term force planning is a mandatory activity not only in industrialized countries but also in Third World countries with larger navies. However, medium naval powers in the developing world frequently find that they are squeezed between requirements that emerge from the planning process and the realities of the financial framework within which they have to operate.

This pattern of planning is applicable in developing countries such as Argentina, Brazil and India, for example, which have institutionalized decision-making on armaments programmes in a way which integrates considerations of the national technology base and budget projections with threat assessments. In India, in particular, naval programmes unfolding today are the product of long-term naval planning dating back to the mid-1960s. Long-term plans, looking at the shape of naval forces up to 15 years ahead, lay out broad objectives. The management of these long-term plans is the subject of a shorter-term 'rolling plan' in which programmes are modified in the light of the immediate national priorities. However, if efforts to lay out long-term plans is standard practice, the experiences of industrialized and developing countries illustrate a gap between the circumstances in different countries. Although in both groups policy involves making choices and establishing priorities, among developed countries budgeting is likely to be predictable enough to allow more cost-effective management of the procurement process. Programmes are meticulously planned several years in advance—including scheduled re-fits and life extension programmes—and integrated closely with both other defence programmes and non-military budget items. There may still be significant changes in short-term plans—most often because of technology failures, cost overruns or changes in budget priorities. However, there is a relatively narrow band within which possible economic performance is likely to fall. Moreover, budgeting takes place against a stable political and strategic backdrop in which dramatic changes in alignment or radical alterations in threat perception are not likely.

This situation does not pertain in developing countries. If major naval powers modernize in a meticulously planned environment where budgets can be predicted at least within certain parameters, this is not true for some countries, where equipment programmes not only have to be responsive to changes in the political and strategic environment; they must also be accommodated by a volatile and vulnerable economy. All governments prefer to avoid spending 'bulges' in which major equipment programmes have to be financed simultaneously, but in some developing countries even rough projections of the total annual revenue available to the government may not be predictable beyond the immediate future, making it difficult to commit funds to future programmes.

The difficulty of matching future procurement programmes with future budget requirements is particularly acute in developing countries. However, this monograph illustrates that this consideration is far from irrelevant for industrialized countries. The cases of Australia and Canada highlight the disruptive impact which budget constraints can have on procurement programmes derived from strategic plans that pay inadequate attention to economic realities.

III. The influence of alliance and alignment

Naval weapon acquisition creates an especially difficult tension within an essentially maritime alliance such as NATO. On the one hand, naval forces play a unique role in sustaining national and international perceptions about the place of a given country within the international system. For island peoples, they are a symbol of self-reliance and independence. On the other hand, however autonomous and independent states would like to appear, alliance membership is an admission by governments that they need each other. The complexity and expense of a naval industry and its attendant infrastructure have been noted. It is difficult for new countries to develop and expensive to maintain in those countries where it already exists. To take alliance membership to its logical conclusion would mean giving up the ability or the aspiration to provide for the national defence without foreign assistance, and for any government this is a difficult step.

The deliberations of NATO members over meeting their future naval requirements have reflected the growing general emphasis on multinational or collaborative weapon programmes. In Europe and the North Atlantic area, members of NATO do not anticipate acting singly, and given the extent of joint operational responsibility and interdependence, a basic lack of military and political trust ought not to be a constraint on collaboration. However, this sense of trust and co-operation does not extend as clearly into the industrial and economic sphere. Among the larger defence producers, the fact of planning for joint military operations has never fed directly into a movement away from national solutions to procurement problems. In the area of the political economy, alliance membership has not in the past been able to solve problems stemming from their respective national interests in maintaining or enhancing economic and technological performance.

The point where governments will have to come to terms with economic and industrial interdependence is being reached even in the naval sphere as a consequence of three related develoments. First, the costs of developing and producing weapons of all kinds have increased with the development of new technologies. All governments have accepted the need to incorporate into the design of modern naval weapon systems a spectrum of technologies that are no longer to be found in one country at the most advanced level. Second, budgetary pressures have either reduced or slowed the growth in defence expenditure in real terms throughout the world since the mid-1980s. Third, political pressures have mounted behind an effort at tackling the grotesque allocation of resources to defence. As illustrated in chapter 4, the development of greater future interdependence between the defence industrial sectors of different countries is certain in the naval sphere.

In considering these issues there will be growing difficulties in isolating naval weapon technologies from other military programmes as a result of the widespread adoption of the 'family' of weapon systems concept. Particular naval air defence systems used as case studies here suggest an emerging pattern of competitive tendering for NATO requirements by consortia of companies based around US or French project leadership rather than a growth in the number of truly multinational defence companies. They have also underlined the purely legal obstacles to genuine trans-Atlantic co-operation. Once it is involved in a defence programme, the Government of the United States is legally obliged to safeguard existing US technological leadership and to promote US leadership in new areas of technology development. However, this broad conclusion from what has been a fairly limited number of cases should not obscure the complexity of the linkages which are developing in the overall sphere of defence production generally. These include the presence of French companies in US-led consortia and vice versa as well as a current lack of clarity concerning the future legal framework for defence contracting after the entry into force of the Single European Act.

Transfers of military technology have not lost their traditional political utility. Arms transfer relationships are interpreted by third parties as unambiguous indicators of a close bilateral relationship. In fact, shifts within an arms transfer relationship are often used as the yardstick by which to measure whether an overall bilateral relationship is improving or deteriorating and carry a weight which is seldom attached to other forms of transaction. Decisions concerning arms programmes can therefore be used by both supplier and recipient governments to convey messages either to one another or to third parties about the nature of their bilateral relationship. That this is true also for naval programmes is perhaps most obvious for the trade within NATO and between the United States and Japan. The location of Japan makes it the country which can most easily impede Soviet naval activity in the Pacific. However, other examples of using transfers of naval systems to sustain the capability of allied naval forces and to complicate the operations of hostile countries have emerged. The recipients of Soviet systems have tended to be countries in parts of the world that the US Navy regards as important operational areas. This applies not only to transfers of ships, but also to submarines and aircraft with their associated naval weapon systems.

Changes in attitude to alliance membership bring direct changes in the weapon acquisition process. Alliances are contracts struck by governments, and the costs of changing the terms of these contracts are not trivial. The redefinition of New Zealand's participation in the ANZUS alliance during 1984–86 was of a different scale from Australia's more limited movement towards a greater regional responsibility. Australia's attitude to regional security has not been set in the context of an alternative to bilateral relations with the United States. On the contrary, it has been worked out in consultation with the United States in the context of increases in Soviet activity in the Pacific and the growing naval capabilities of ASEAN countries to the north of Australia. This kind of reassessment of alliance membership has not interrupted the US–Australia defence relationship. For New Zealand, being effectively outside the framework of the alliance has meant that Wellington finds its access to certain types of weapon system restricted. Conversely, US permission for the transfer of nuclear-propulsion technologies to Canada from the United Kingdom would not have been forthcoming had Canada not been an ally. However, it is equally true that refusing permission would have been difficult to explain to the satisfaction of Ottawa. Alliance membership sets parameters on the behaviour of all members, including the strongest.

Naval arms transfers can become sources of frustration if not outright tension within an alliance. In 1988 the United States has voiced its frustration at not being able to meet Argentine requests for assistance, in particular with spare parts for US aircraft, because of British objections to arms sales to Argentina in advance of a settlement of the Falklands/Malvinas question. In the case of Australia, when Canberra objected to US sales to other South-East Asian countries this was mentioned as part of a much more complex re-evaluation by Australia of its attitudes towards the utility of the ANZUS alliance in contingencies short of a major conflict.

Countries which make their procurement policy outside of an alliance framework face even more difficult choices between different types of equipment programme because they have no opportunity for agreements over a division of labour. In all of the cases where there has been an absence of formal alliance relationships, this has stimulated a long-term interest in the development of a broad defence industrial base. There has been a heavy investment in shipyards and construction, and a degree of domestic production, usually under licence, is a condition of signing arms import agreements. However, no neutral or non-aligned country has been able to pursue defence industrialization without considerable foreign assistance.

New ships cost a great deal of money, especially if they are purchased from countries which are not in a position to supply them as part of a wider military aid or assistance agreement of some sort. The capital costs of naval programmes are so high partly because they involve much more than the acquisition of ships. Shore facilities, logistic support and training personnel require a major investment of human and material resources.

It is doubtful whether any Third World country will be able to develop a fully autonomous shipbuilding and naval equipment industry and among industrialized countries only Japan could do so (if the Japanese Government had sufficient political will). There are growing questions about whether traditional shipbuilding countries, including the superpowers, will be able to sustain their existing capabilities beyond the near term.[1] As noted above, there is a new maritime environment emerging in which major powers cannot control events to the extent that was once possible. To test the implications of this new environment by military means cannot meet any of the interests of the affected parties. This means that these issues are increasingly becoming questions for political dialogue both bilaterally and at the regional level. Arms control should be a component of that dialogue but in the sphere of interest of this book—the transfer of conventional arms for use at sea—the political climate is not yet right for arms control.

IV. Prospects for naval arms control

Naval arms control is not a central issue in this monograph, reflecting the twin realities that limiting conventional arms transfers and limiting the size of naval forces are not currently on the international arms control agenda.

In 1988 and 1989 there were several new developments relevant to the control of international transfers of conventional armaments. In 1988 there were statements in several multinational forums, notably at the United Nations Third Special Session on Disarmament (SSOD III), which pointed to the need for new initiatives in the area of arms trade control. More important, progress towards a more meaningful conventional arms control framework in Europe after the closure of the fruitless and sterile 14-year Mutual and Balanced Force Reduction (MBFR) talks is likely to stimulate efforts to regulate the production of conventional military equipment.

These developments, were they to occur, would clearly affect all arms exports. However, this is not likely to have an impact in the area of naval systems for several reasons.

First, naval forces are explicitly excluded from the mandate currently agreed by negotiators at the Negotiation on Conventional Armed Forces in Europe (CFE). This situation may change, especially since the Soviet Union has consistently argued that excluding naval forces distorts the overall arms control process: although naval forces are not *restricted* to operating in the European theatre, at the same time they are clearly *relevant* to military operations in the European theatre.[2]

Second, as discussed in chapter 2, naval weapon systems represent a very important component of the overall arms exports, in particular from West European suppliers. Consequently there is likely to be even greater resistance to arms export control in the naval area than elsewhere from important interest groups both in industry and within ministries responsible for defence production.

Third, the concept of arms trade control meets with considerable resistance from recipient countries in the Third World who reject the idea that they are currently overarmed.[3] This is a widespread and firmly held conviction based on observation. In 1987 industrialized countries were responsible for 83 per cent of military expenditure.[4] Moreover, as noted in chapter 3, developing countries feel particularly vulnerable in the area of their naval forces because of recent trends in maritime development.

As noted in chapter 3, the long negotiations at the Third United Nations Law of the Sea Conference illustrated that more and more coastal states are interested in the extension of their national jurisdiction over both the sea bed (and its subsoil) and over sea areas adjacent to their coastlines. At the same time, the Law of the Sea as it exists does award coastal states sovereignty over these sea territories. In thinking about these issues of maritime jurisdiction many coastal states have become very aware of how limited their own national capacities for effective surveillance and action at sea are.

These factors of newly awakened sensitivity to the importance of the sea, the lack of a global legal framework for regulating maritime activity and heightened awareness of maritime vulnerability were at the root of many of the programmes discussed in this monograph. It is therefore reasonable to conclude that effective limits over transfers of naval conventional armaments will not occur before there is progress in two areas.

Establishing an effective global framework regulating maritime activity may be a prerequisite for naval arms control in that without such a framework coastal states will feel obliged to acquire the means to protect their own interests. However, a satisfactory global regime must accommodate the conflicting interests of coastal states looking to expand their maritime jurisdiction and other sea users who insist on preserving their traditional freedom of the seas. If they cannot persuade sea users to give up this traditional freedom—and a persuasive argument is not easy to imagine—then coastal states with sufficient power and resources may feel compelled simply to assert their jurisdiction as best they can.

Below the global level, some countries may be prepared to consider regional agreements that heighten crisis stability and reduce the danger of war—perhaps broadly modelled on the 1972 Incidents at

Sea Agreement between the United States and the Soviet Union. Such measures, worthwhile though they may be, would not, however, settle the many outstanding disputes over the demarcation of maritime boundaries especially in areas where several states share short stretches of coastline or are located around the edges of semi-enclosed seas, such as the South China Sea, the Yellow Sea and the Sea of Japan.

These problems of conflicting interests are a major obstacle to arms control. Another is the growing imbalance of power between major navies—in particular the US Navy—and the navies of other coastal states. As is argued in several places in this monograph, the maritime vulnerability of even the more advanced Third World countries in the face of the naval forces of larger powers was demonstrated in the 1982 Falklands/Malvinas War and by the engagements between the US Navy and Iranian forces in the Persian Gulf in 1988. In the eyes of most of the larger developing countries, conventional arms transfer control would simply cement this imbalance permanently and reinforce the already considerable ability of developed countries to dictate the terms of political development throughout the world.

Finally, one might add that the hostility of arms recipients to conventional arms transfer control is likely to be greatest in some specific areas of interest to the major naval powers. As noted in chapter 3 and in the cases of Argentina, Brazil, Canada and India, the spread of more advanced technologies—such as nuclear-powered submarines and the next generation of supersonic anti-ship missiles—would complicate the maritime operations of major naval powers. For this reason there may be an interest in preventing the spread of such systems among some of the important arms suppliers including the US. However, it is precisely these systems which developing countries see as 'equalizers' in the overall naval force balance.

It is argued here that arms control is less likely to play a role in the spread of these systems than the sheer technical complexity and expense of developing and maintaining the capability to use them effectively. It will take smaller naval powers a considerable time to develop larger forces. If, during this period, naval forces were to be included in the broader conventional arms control process, whether in the regional context of European security or in the bilateral US–Soviet framework, and if this process led in turn to significant force reductions in major power navies, then the climate for negotiating limits to the growth of smaller navies may improve. Without such a development the prospects for controlling the spread of naval technology remain remote.

Notes and references

1. Fieldhouse, R. and Taoka, S., SIPRI, *Superpowers at Sea: An Assessment of the Naval Arms Race*, Strategic Issue Papers (Oxford University Press: Oxford, 1989).

2. The issue of naval arms control is thoroughly developed in Fieldhouse, R. (ed.), *Security At Sea: Naval Forces and Arms Control* (Oxford University Press: Oxford, 1990).

3. This perspective is forcefully argued in Ohlson, T., SIPRI, *Arms Transfer Limitations and Third World Security* (Oxford University Press: Oxford, 1989).

4. *World Military Expenditures and Arms Transfers 1988* (Arms Control and Disarmament Agency: Washington, DC, 1989), p. 1.

Appendices

Appendix 1. Second-hand ships transferred 1947–88

Seller/ Buyer	Designation	Description	Years of delivery	No. deliv.	Comments
Algeria					
Nicaragua	Yevgenia class	MSC	1983–84	2	Delivered via Cuba
Argentina					
Paraguay	Bouchard class	MSC	1964–68	3	Ex-*Argentina*
Australia					
China	Bathurst class	MSO	1947	1	
Indonesia	Attack class	Patrol craft	1972–74	2	
Indonesia	Attack class	Patrol craft	1981–83	3	
N. Zealand	Bathurst class	MSO	1951–52	4	Scrapped 1967, 1968 and 1976
Belgium					
Norway	Adjutant class	MSC	1966	3	Exchanged for 2 MSOs
Taiwan	Adjutant class	MSC	1968–69	7	1 bought for spare parts; built in USA 1953–55
Canada					
Belgium	Algerine class	MSO	1959	2	Built in the UK
China	Castle class	Frigate	1947	2	Originally lend-lease; bought in 1959
Norway	River class	Patrol craft	1956	3	Modernized 1952; both scrapped 1966
Peru	River class	Patrol craft	1947	2	
Taiwan	Castle class	Frigate	1947–51	2	Sold as merchant ship to China 1947; armed by Taiwan in 1951; scrapped by 1967
Turkey	Bay class	Frigate	1958	4	
China					
Angola	Shanghai class	Patrol craft	1975	1	

Country	Class	Type	Years	Number	Notes
Bangladesh	Shanghai class	Patrol craft	1980–82	8	Option on more
Bangladesh	Hegu class	FAC	1983	4	
Bangladesh	P-4 class	FAC	1982–83	4	
Cameroon	Shanghai class	Patrol craft	1975–76	2	
Cape Verde	Shanghai class	Patrol craft	1975	2	
Egypt	Romeo class	Submarine	1984–86	2	Followed by licensed production of 8
Korea N.	Romeo class	Submarine	1973–75	7	
Korea N.	Hainan class	Patrol craft	1974–78	6	
Korea N.	Shanghai class	Patrol craft	1967–78	23	Total number unconfirmed
Korea N.	Huangfen class	FAC	1979–83	2	Chinese-built Osa class FACs
Pakistan	Hegu class	FAC	1980–81	4	
Pakistan	Huangfen class	FAC	1983–84	4	Chinese-built Osa class FACs
Sierre Leone	Shanghai class	Patrol craft	1973	3	
Sri Lanka	Shanghai class	Patrol craft	1971–72	5	
Tanzania	Huchuan class	Hydrofoil FAC	1975	4	
Tanzania	Shanghai class	Patrol craft	1971–72	6	
Tunisia	Shanghai class	Patrol craft	1976–77	2	
Vietnam	Wusung class	MSO	1973–74	1	Unconfirmed
Zaire	Shanghai class	Patrol craft	1975–78	4	
Zaire	Huchuan class	Hydrofoil FAC	1978–79	4	
Denmark					
S. Africa	Tafelberg class	Support ship	1965	1	Modified for naval use in South Africa
France					
FRG	Type 319	MSO	1955–57	5	
Greece	Amalthee class	Support ship	1976	1	Built in Germany, taken over by French
Japan	LSM type	Landing craft	1958	1	
Kampuchea	LSIL type	Landing craft	1956–57	2	US ships bought by France in 1951

Seller/ Buyer	Designation	Description	Years of delivery	No. deliv.	Comments
Kampuchea	PC-452 type	Patrol craft	1954–56	2	US ships bought by France in 1951
Kampuchea	LCU-501 class	Landing craft	1954–57	4	US ships bought by France
Madagascar	EDIC/EDA type	Landing craft	1984–85	2	
Morocco	Sirius class	MSC/PC	1974	1	On loan
Morocco	Annamite class	Corvette	1961	1	Returned to France 1967
Morocco	River class	Frigate	1964	1	Built in UK
Morocco	PC-452 type	Patrol craft	1960	1	Built in USA; returned to France 1964
Senegal	EDIC/EDA type	Landing craft	1973–74	1	
Senegal	SC type	Patrol craft	1961	1	
Seychelles	Sirius class	MSC/PC	1978–79	1	Minesweeping gear removed
Seychelles	Sirius class	MSC/PC	1979	1	Gift
Tunisia	Annamite class	Corvette	1959	1	
Uruguay	Adjutant class	MSC	1969	1	US ship bought by France
Uruguay	Vigilante class	Patrol craft	1979–81	1	
USA	Independence class	Support ship	1960–63	2	Handed back by France
Vietnam (S.)	LCU 1466 class	Landing craft	1954	4	US ship given by France as reparations
Vietnam (S.)	LCU-501 class	Landing craft	1954	1	US ship given by France as reparations
Vietnam (S.)	LSIL type	Landing craft	1955–56	7	US ship given by France as reparations
Vietnam (S.)	LSM type	Landing craft	1954–55	2	US ship given by France as reparations
Vietnam (S.)	LSSL type	Support ship	1954–57	2	US ship given by France as reparations
Vietnam (S.)	PC-452 type	Patrol craft	1954–56	5	US ship given by France as reparations
Vietnam (S.)	SC type	Patrol craft	1954	2	US ship given by France as reparations
Vietnam (S.)	YMS class	MSC	1954	3	Ex-US MSC of 232t

FR Germany

Greece	Jaguar class	FAC	1975–77	6	
Greece	Jaguar class	FAC	1976–77	6	3 for spares
Greece	Rhein class	Support ship	1975	1	
Greece	Rhein class	Support ship	1976–76	1	
Greece	Fletcher class	Destroyer	1979–81	2	Transferred for spares
Greece	Fletcher class	Destroyer	1980–82	1	NATO aid in addition to 6 in service
Sweden	Branmaren	Tanker	1971–72	1	
Tunisia	Le Fougueux class	Patrol craft	1969	1	Built in France
Turkey	Type 763	Support ship	1977	1	
Turkey	Mercure class	MSC	1974–79	7	Built in France
Turkey	Rhein class	Support ship	1975–83	2	
Turkey	Koeln class	Frigate	1982–84	3	NATO aid
Turkey	LST 1-510	Landing ship	1972	2	
Turkey	Jaguar class	FAC	1974–76	10	3 for spare parts
Turkey	Angeln class	Support ship	1972–75	2	Built in France
Turkey	Zobel class	FAC	1984–84	6	
Norway	Type XXIII	Submarine	1950–50	1	Transferred via UK
Norway	Type VIIC	Submarine	1950–50	3	
Norway	Type 1940	MSC	1948–48	13	Transferred via UK
Norway	S-Boat type	FAC	1947–47	7	1 retransferred to Denmark

Iceland

Norway	Thor class	OPV	1984	1	Built in Denmark 1951

India

Bangladesh	Abhay class	Patrol craft	1972–74	2	
Mauritius	Abhay class	Patrol craft	1974–74	1	

Seller/ Buyer	Designation	Description	Years of delivery	No. deliv.	Comments
Israel					
Chile	Reshef class	FAC	1979–81	2	Armed with Gabriel ShShMs
Sri Lanka	Dabur class	FAC	1984	2	
Sri Lanka	River class	Frigate	1958–59	2	Possibly Taiwanese Hai Ou class
Italy					
France	Eritrea class	Frigate	1948	1	Acquired under Peace Treaty of 1948
France	Orani class	Frigate	1948–61	4	Acquired under Peace Treaty of 1948
France	A. Regolo class	Destroyer	1948	2	Acquired under the Peace Treaty of 1948
FRG	Zotti class	Tanker	1960	2	Converted 1960-61 from civilian use
Yugoslavia	Ramb III	Cruiser/minelayer	1952	1	Used as cadet training ship
Korea N.					
Benin	P-4 class	FAC	1978–79	2	
Guyana	Sin Hung class	FAC	1980	5	Probably without torpedoes
Nicaragua	Sin Hung class	FAC	1983	2	Unconfirmed
Libya					
Malta	Type A/B	Patrol craft	1978	3	Built in the UK 1967-69
Netherlands					
Argentina	Colossus class	Aircraft-carrier	1968	1	Ex-UK
Ethiopia	Dokkum class	MSC	1973	1	
Indonesia	Bathurst class	MSO	1948–50	4	Originally Australian; received from Dutch; 1 sunk 1958; 3 scrapped 1968–69
Indonesia	Djember class	MSC	1951	4	Built in Indonesia before independence; deleted 1974
Indonesia	N class	Destroyer	1950–51	1	Ex-UK; deleted 1961
Indonesia	Sawega type	Support ship	1956–57	2	

	Class	Type	Years	No.	Remarks
Indonesia	LCI type	Landing craft	1949–50	5	Ex-US ships transferred on independence
Indonesia	V. Speijk class	Frigate	1986–88	4	Request for 2 more
Kuwait	Alkmaar class	Minehunter	1987–88	2	
Oman	Dokkum class	MSC	1974–74	2	
Peru	Holland class	Destroyer	1978	1	
Peru	De Ruyter class	Cruiser	1973–76	2	Terrier SAMs returned to USA before sale
Peru	Friesland class	Destroyer	1980–82	6	
Singapore	Endeavour	Patrol craft	1969–70	1	Patrol craft built in FRG 1955; gun-armed
Norway					
Belgium	Aggressive class	MSO	1966	2	In addition to 18 taken over directly from German Navy
Denmark	S-Boat type	FAC	1951	6	
France	Papenoo class	Tanker	1971	2	
France	La Charente	Tanker	1964	1	Ex-*Norway*
Greece	Tjeld class	FAC	1967	5	
Korea S.	Birk type	Tanker	1951	2	Tankers built in 1951
Turkey	LSM type	Landing craft	1960	2	Built in USA
Pakistan					
Bangladesh	Town class	Patrol craft	1978	1	Sunk in 1971 Indo-Pakistani War; salvaged by Bangladeshi Navy
Poland					
Indonesia	Skory class	Destroyer	1958–59	4	All scrapped by 1973
Indonesia	Whiskey class	Submarine	1958–59	2	Scrapped 1972
Syria	Polnocny class	Landing ship	1983–85	3	

Seller/ Buyer	Designation	Description	Years of delivery	No. deliv.	Comments
Portugal					
Angola	Flower class	Frigate	1975	1	Handed over at independence
Angola	Argos class	Patrol craft	1975	5	Handed over at independence
Angola	Alfange class	Landing craft	1975	1	Handed over at independence
Mozambique	Alfange class	Landing craft	1975	1	Handed over at independence
Pakistan	Daphne class	Submarine	1975	1	
S. Africa					
Israel	Bat Sheva	Support ship	1967–68	1	
Singapore					
Sri Lanka	Abheetha class	Support ship	1984	3	Used as command ships
Sri Lanka	Mahawele class	Support ship	1984	3	Called surveillance tenders; supplier unconfirmed
Sweden					
Chile	Tre Kronor class	Cruiser	1970–71	1	
Guatemala	Jagaren class	Patrol craft	1958–59	1	
Syria					
Egypt	P-4 class	FAC	1969–70	4	
Turkey					
Libya	C-107 class	Landing craft	1979–81	20	Unconfirmed orders for up to 30 more
UK					
Argentina	River class	Frigate	1948	1	Transferred to Coast Guard 1961
Argentina	Colossus class	Aircraft-carrier	1958–59	1	Modernized 1952–53; scrapped 1971
Argentina	Ton class	Minehunter	1968–69	6	Converted into minehunter
Australia	Majestic class	Aircraft-carrier	1949–56	1	*HMS Majestic*; renamed *Melbourne*; modernized 1967–68, 1971–73 and 1976; transferred to reserves 1982

Country	Class	Type	Year	Number	Notes
Australia	Colossus class	Aircraft-carrier	1951–52	1	*Vengeance*; returned to UK 1955
Australia	Ton class	Minehunter	1961–62	6	Converted to minehunters 1967–70
Australia	Daring class	Destroyer	1964	1	On loan until 1972 when bought
Bangladesh	Salisbury class	Frigate	1976	1	
Bangladesh	Leopard class	Frigate	1977–78	1	
Bangladesh	Leopard class	Frigate	1981–82	1	
Belgium	Algerine class	MSO	1949–53	6	All out of service by late 1960s
Belgium	Kamina class	Support ship	1950	1	Originally built in Belgium for Poland; seized by Germany, then UK
Burma	Algerine class	MSO	1957–58	1	Fitted for mine-laying; built in Canada 1943–44
Canada	Majestic class	Aircraft-carrier	1948–52	2	*Magnificent* and *Powerful*
Chile	County class	Destroyer	1981–84	2	
Chile	County class	Destroyer	1986–87	1	*HMS Glamorgan*
Chile	County class	Destroyer	1987	1	*HMS Fife*
China	Flower class	Frigate	1949	3	Originally transferred as merchant ship and armed in China; in service until 1979
Denmark	Hunt class	Frigate	1953	3	Out of service by mid-1960s
Ecuador	Hunt class	Frigate	1955	2	
Egypt	Fairmile-D type	FAC	1950–51	2	
Egypt	Bangor type	MSO/Corvette	1949–50	3	
Egypt	Flower class	Frigate	1948–49	2	
Egypt	River class	Frigate	1947–48	3	
Egypt	Hunt class	Frigate	1949–51	2	
Egypt	Black Swan class	Frigate	1948–49	1	
Egypt	Cavendish class	Destroyer	1954–55	2	Also designated Z class
Finland	Bay class	Frigate	1962	1	Scrapped 1975
France	Colossus class	Aircraft-carrier	1946	1	*Arromanches*; withdrawn in 1974

Seller/ Buyer	Designation	Description	Years of delivery	No. deliv.	Comments
France	S-Boat type	FAC	1951	4	Loaned by Royal Navy for ASW training; out of service 1952–1961
FRG	Black Swan class	Frigate	1957–59	4	
FRG	Hunt class	Frigate	1957–59	3	German designation Frankenland
FRG	Powell class	Support ship	1959	1	Build in Canada as minesweepers
FRG	Isles class	Support ship	1956	2	Loaned
Ghana	Ton class	Minehunter	1962	1	
Greece	U class	Submarine	1943–46	6	On loan; sunk as targets 1957
India	R-Q class	Destroyer	1948–50	3	
India	Hunt class	Frigate	1952–53	3	
India	Ton class	Minehunter	1956	4	
India	Fiji class	Cruiser	1954–57	1	
India	Majestic class	Aircraft-carrier	1957–61	1	HMS Hercules; renamed *Vikrant*;
India	Hermes class	Aircraft-carrier	1986–87	1	Deal worth approx $74 m
Indonesia	Pladju type	Support ship	1957–58	1	Gun-armed oiler purchased from Singapore
Indonesia	Hecla class	OPV	1985–86	1	
Indonesia	Tribal class	Frigate	1984–86	3	
Iran	Algerine class	MSO	1948–49	1	
Iran	Loch class	Frigate	1948–49	1	Decommissioned 1972
Iran	Battle class	Destroyer	1964–67	1	Refitted before delivery
S. Africa	W class	Destroyer	1950–56	3	Refitted 1962–66 to carry Wasp helicopter; two more cancelled 1956
Ireland	Flower class	Frigate	1945–46	3	
Ireland	Ton class	Minehunter	1970–71	2	For fishery protection
Ireland	Peacock class	OPV	1988	2	Deal worth $12 m
Israel	Flower class	Frigate	1949–50	2	
Israel	River class	Frigate	1949–51	3	First 2 resold to Ceylon 1959

Country	Class	Type	Date	No.	Notes
Israel	Cavendish class	Destroyer	1954–55	2	Ex-RN Z class destroyers; *Eilat* sunk by Egyptian Styx ShShMs 1967
Israel	Hunt class	Frigate	1955–56	1	Israeli designation: *Haifa*
Israel	S class	Submarine	1956–60	2	
Israel	T class	Submarine	1964–68	3	
Italy	Flower class	Frigate	1949	1	
Kenya	Ford class	Patrol craft	1964	1	Loaned in 1964; bought in 1967
Malaysia	Ham class	MSC	1957–59	4	Scrapped 1966–67
Malaysia	Ton class	Minehunter	1959–60	1	Scrapped 1980
Malaysia	Loch class	Frigate	1962–64	1	Refitted before delivery; scrapped 1977
Malaysia	Ton class	Minehunter	1962–66	5	Refitted before delivery; scrapped by 1982
Malaysia	Ham class	MSC	1964–66	2	Sweeping gear replaced with guns before delivery
Malaysia	Ton class	Minehunter	1968–69	1	In addition to 6 in service; converted to survey ship before delivery
Malaysia	Mermaid class	Frigate	1975–1977	1	
Netherlands	Colossus class	Aircraft-carrier	1948	1	*Venerable*, purchased and fully modernized in the 1950s; Dutch designation *Karel Doorman* sold to Argentina 1968
Nigeria	LCT-4 type	Landing ship	1959–59	1	
Nigeria	Ford class	Patrol craft	1960–68	7	Used as survey ship
Nigeria	Bulldog class	OPV	1973–76	2	
Norway	Hunt class	Frigate	1952	2	Originally lent; bought 1956
N. Zealand	Loch class	Frigate	1947–49	6	2 scrapped in 1961, 2 in 1965, 2 in 1966
N. Zealand	Bellona class	Cruiser	1953–56	1	Lent to New Zealand with modernized armament; UK paid for maintenance; scrapped 1968
N. Zealand	Whitby class	Frigate	1965–66	1	On loan 1966–71

Seller/ Buyer	Designation	Description	Years of delivery	No. deliv.	Comments
N. Zealand	Leander class	Frigate	1981–84	2	*Wellington* and *Southland* refitted with Wasp helicopter and Seacat ShAMs
Pakistan	Bellona class	Cruiser	1955–56	1	
Pakistan	O class	Destroyer	1949–51	3	
Pakistan	Battle class	Destroyer	1955–57	1	
Pakistan	CR class	Destroyer	1956–58	3	
Pakistan	County class	Destroyer	1981–82	1	
Pakistan	Leander class	Frigate	1988	2	Ex-Royal Navy ships *HMS Diomede* and *HMS Apollo*
Peru	Fiji class	Cruiser	1959–60	2	Modernization of 1 ship 1955–56
Peru	Daring class	Destroyer	1969	2	Refitted 1970–73; helicopter landing pad added; armed with MM-38 Exocet ShShMs
Peru	Fiji class	Cruiser	1955–60	2	Modernized 1955–56 and sold to Peru
Portugal	River class	MSO	1948–49	2	Transferred May 1949; refitted 1959 with ASW equipment
Portugal	Bay class	Frigate	1959–61	4	
S. Africa	Ton class	Minehunter	1955–59	8	
S. Africa	Algerine class	MSO	1947	2	
S. Africa	Ford class	Patrol craft	1954–55	2	Refitted 1961–62
Sri Lanka	Ford class	Patrol craft	1955	1	
Sri Lanka	Algerine class	MSO	1958–58	1	In addition to 2 acquired 1949
Turkey	Auk class	Minelayer	1946–47	8	Lend-leased to UK
Turkey	Bathurst class	MSO	1945–46	6	Built in Australia
Turkey	Milne class	Destroyer	1953–59	4	
Turkey	Bangor class	MSC	1957–58	10	Built in Canada
Turkey	LCM-9 type	Landing craft	1966–67	6	
USA	Lyness class	Support ship	1980–83	3	First 2 ships sold for $37 m, 3rd for $10.5 m

Yemen S.	Ham class	MSC	1967	3	
Yugoslavia	W class	Destroyer	1955–56	2	Refitted in Yugoslavia

USA

Argentina	LSM type	Landing craft	1947–49	2	1 deleted 1971
Argentina	Flower class	Frigate	1948	1	
Argentina	LST 511-1152	Landing ship	1950–52	14	4 deleted 1958–60; five in 1964
Argentina	Brooklyn class	Cruiser	1951	2	
Argentina	Balao class	Submarine	1960	2	
Argentina	Fletcher class	Destroyer	1961	3	All deleted 1982
Argentina	Ashland class	Landing ship	1970	1	Built during WW II
Argentina	Fletcher class	Destroyer	1971	2	
Argentina	Balao class	Submarine	1971	2	
Argentina	LSM type	Landing craft	1971	4	
Argentina	Sumner class	Destroyer	1972	2	
Argentina	Gearing class	Destroyer	1973	1	MM-38 Exocet added 1977–78
Argentina	Sumner class	Destroyer	1974	2	MM-38 Exocet added on one 1977–78
Brazil	Cannon class	Frigate	1944–45	8	Lend-and-Lease programme
Brazil	Brooklyn class	Cruiser	1951	2	1 ship sold 1975
Brazil	Gato class	Submarine	1957	2	Refitted before transfer; 1 returned 1969. 1 deleted 1968
Brazil	Fletcher class	Destroyer	1959–61	4	Loaned; 2 scrapped 1978; 2 still in service
Brazil	YMS class	MSC	1960–63	4	
Brazil	LST 511-1152	Landing ship	1962	1	Loaned 1962; bought 1977
Brazil	Fletcher class	Destroyer	1967–73	3	Loaned
Brazil	Parish class	Landing ship	1973	1	
Brazil	Gearing class	Destroyer	1973	2	Armed with ASROC and Wasp helicopter
Brazil	LCU-501 class	Landing craft	1973	1	Loaned May 1971, bought 1973

Seller/Buyer	Designation	Description	Years of delivery	No. deliv.	Comments
Brazil	Guppy-2 class	Submarine	1972–73	7	
Brazil	Sumner class	Destroyer	1972–73	5	1 armed with Seacat SAMs
Burma	Admirable class	MSO	1966–67	1	Minesweeping gear removed
Burma	PCE-827 class	Corvette	1964–65	1	
Canada	Tench class	Submarine	1967–68	1	Returned to US Navy 1974 for scrap
Canada	Balao class	Submarine	1960–61	1	Returned to US Navy 1969 for scrap
Chile	LSM type	Landing craft	1947–50	4	
Chile	LCI type	Landing craft	1949–50	6	
Chile	Brooklyn class	Cruiser	1951	2	
Chile	LSM type	Landing craft	1960	1	
Chile	Balao class	Submarine	1961	2	
Chile	PC-1638 type	Patrol craft	1969–70	1	Prior to licensed production of 3
Chile	Sumner class	Destroyer	1974	2	
China	Adjutant class	MSC	1975–76	3	Used as PC in China
China	LST 511-1152	Landing ship	1946	15	
China	YO type	Support ship	1945–46	3	
China	FS-330 type	Support ship	1945–46	8	
China	LST 1-510	Landing ship	1946	1	
China	LSIL type	Landing craft	1946	6	Some used as minesweepers
China	Gordy class	Frigate	1954–55	4	ShShMs fitted 1971–74
Colombia	Asheville class	Frigate	1982–84	2	Leased by Navy; for coastal patrol
Colombia	Tacoma class	Frigate	1947–53	3	All discarded late 1960s
Colombia	Dealey class	Frigate	1972	1	
Colombia	Buckley class	Frigate	1965–69	4	
Colombia	Sumner class	Destroyer	1972–73	2	
Colombia	Fletcher class	Destroyer	1960	1	
Denmark	LSM type	Landing craft	1953–54	3	

Country	Class/Type	Ship type	Years	No.	Notes
Denmark	YO type	Support ship	1960–62	2	
Dominica	Osprey class	Patrol craft	1982	3	
Domin. Rep.	Admirable class	MSO	1964–65	2	
Domin. Rep.	PGM-71 class	Patrol craft	1965–66	1	
Netherlands	Cannon class	Frigate	1950–51	6	Bought with MAP funds 1950
Domin. Rep.	LSM type	Landing craft	1957–58	1	
Domin. Rep.	Cohoes class	OPV	1976	3	
Domin. Rep.	PCE-827 class	Corvette	1960	2	
Ecuador	LST 511-1152	Landing ship	1977	1	Transferred after extensive refit
Ecuador	LSM type	Landing craft	1958–59	2	
Ecuador	PT type	FAC	1947–49	1	
Ecuador	Gearing class	Destroyer	1978	1	Purchased 1978
Ecuador	Ch Lawrence class.	Destroyer	1967	1	
Egypt	YMS class	MSC	1948–50	8	2 retransferred to Algeria in 1962
France	LSD-13 class	AALS	1952	1	Scrapped in 1969
France	LST 1-510	Landing ship	1949–51	10	Out of service by 1970
France	Cannon class	Frigate	1950–52	8	
France	LSM type	Landing craft	1954	1	Later resold to Japan
France	Independence class	Support ship	1950–53	2	Returned to US Navy in 1960 and 1963
France	SC type	Patrol craft	1950–51	2	Later retransferred to S. Vietnam
France	LCU 1466 class	Landing craft	1950–51	4	Later retransferred to S. Vietnam
France	LCU-501 class	Landing craft	1950–51	1	Later retransferred to S. Vietnam
France	LSIL type	Landing craft	1950–53	7	Later retransferred to S. Vietnam
France	LSM type	Landing craft	1953–54	2	Later retransferred to S. Vietnam
France	LSSL type	Support ship	1950–51	2	Later retransferred to S. Vietnam
France	PC-452 type	Patrol craft	1950–51	5	Later retransferred to S. Vietnam
FRG	LST 511-1152	Landing ship	1960–61	7	3 converted to minelayers, 2 converted to maintenance ships and 2 scrapped

Seller/ Buyer	Designation	Description	Years of delivery	No. deliv.	Comments
FRG	R-Boat type	MCM	1956–57	24	Originally built in Germany, taken by US Navy and returned to FRG
FRG	Harle class	Tanker	1956	2	Originally built in Germany and Norway, taken by US Navy and returned to FRG
FRG	Fletcher class	Destroyer	1956–60	6	Leased
FRG	LSM type	Landing craft	1957–58	4	
FRG	LSMR type	Landing ship	1957–58	2	Total cost with 2 LSM type: $6 m
Greece	Parish class	Landing ship	1975–76	2	Ex-*USS Terrell County* and *Whitfield County*
Greece	Gearing class	Destroyer	1976–78	3	
Greece	Gearing class	Destroyer	1979–81	4	
Greece	Parish class	Landing ship	1977	2	Purchased 1977
Greece	Adjutant class	MSC	1969	5	Supplied to Belgium under MDAP, all built in USA 1954; returned to USA and transferred to Greece
Greece	Algerine class	MSO	1947	5	Formerly employed as corvettes
Greece	LSM type	Landing craft	1958	5	
Greece	LSM type	Landing craft	1951	2	Converted in USA into all-purpose MSO
Greece	Tench class	Submarine	1973	1	Converted into Guppy III class 1961–62
Greece	Balao class	Submarine	1964–72	2	Loaned by USA in 1964 under MAP, purchased Apr. 1976 for training only
Greece	Gato class	Submarine	1957–58	2	Loaned under MDAP; 1 purchased Apr 1976 for spares to keep other submarines operational
Greece	Cannon class	Frigate	1951	4	Refitted with torpedoes
Greece	Gearing class	Destroyer	1971–81	8	Modernized 1960 and 1976–77; 3 purchased 1978
Greece	Sumner class	Destroyer	1971	1	Transferred after modernization 1971

Country	Class	Type	Years	No.	Notes
Greece	Fletcher class	Destroyer	1959–82	9	Transferred under MAP all loans renewed Mar 1970 and purchased 1977; 2 more acquired for spares after being used as targets
Indonesia	PC-452 type	Patrol craft	1955–60	5	Ex-US WWII submarine-chasers
Indonesia	LST 511-1152	Landing ship	1959–60	1	
Indonesia	LST 511-1152	Landing ship	1960–61	5	
Indonesia	LST 511-1152	Landing ship	1969–71	3	In addition to 7 received earlier
Indonesia	Bluebird class	MSC	1970–71	6	Deleted 1976
Indonesia	Arcadia class	Support ship	1970–71	1	Ex-*USS Tidewater*, renamed *Duma*
Indonesia	LST 511-1152	Landing ship	1971	1	Leased after service in Viet Nam War
Indonesia	Claud Jones class	Frigate	1972–74	4	
Iran	Amphion class	Support ship	1970–71	1	Transferred on loan 1971; purchased 1977
Iran	LSIL type	Landing craft	1956–58	2	
Iran	LCU 1466 class	Landing craft	1963–64	1	
Iran	Sumner class	Destroyer	1970–74	2	
Iran	Gearing class	Destroyer	1974–77	2	Bought for spares at scrap value
Israel	LSM type	Landing craft	1971–72	3	
Italy	Tang class	Submarine	1973–74	2	
Italy	LSM type	Landing craft	1952	30	
Italy	LSIL type	Landing craft	1951	6	
Italy	LST-1173 class	Landing ship	1971–72	2	
Italy	Gato class	Submarine	1951–55	2	
Italy	Fletcher class	Destroyer	1969–70	2	
Italy	Cannon class	Frigate	1951	3	
Italy	Benson class	Frigate	1951	2	
Italy	Balao class	Submarine	1960–72	5	Transferred in 3 batches
Italy	Aggressive class	MSO	1956–57	4	

Seller/Buyer	Designation	Description	Years of delivery	No. deliv.	Comments
Italy	Adjutant class	MSC	1953–54	18	Prior to licenced production
Italy	Andromeda class	Support ship	1961–62	1	
Italy	LST 1-510	Landing ship	1962–62	1	Used as transport ship
Italy	YO type	Support ship	1959	1	
Japan	LST 1-510	Landing ship	1961	3	MAP; two deleted 1974–75; last back to USA for resale to the Philippines
Japan	Asheville class	Frigate	1952–53	13	
Japan	Fletcher class	Destroyer	1958–59	2	
Japan	Gato class	Submarine	1954–55	1	MAP; scrapped 1966
Kampuchea	LCU 1466 class	Landing craft	1968–69	1	Seller unconfirmed
Kampuchea	PGM-39 class	Patrol craft	1964	1	Military aid
Kampuchea	LCU-501 class	Landing craft	1961–73	5	Seller unconfirmed
Korea N.	Artillerist class	Patrol craft	1953–54	2	
Korea S.	Gearing class	Destroyer	1979–81	2	In addition to 5 in service
Korea S.	Asheville class	Frigate	1950–52	5	Built 1943–45; deleted 1974; 2 more transferred in 1969 for cannibalization
Korea S.	Gearing class	Destroyer	1975–78	3	
Korea S.	Tonti class	Tanker	1981–82	1	
Korea S.	Asheville class	Patrol craft	1971–72	1	Standard ShShMs fitted in 1975–76
Korea S.	Auk class	Minelayer	1962–67	3	Minesweeping gear replaced with 76 mm gun; built 1941–44; deleted 1982
Korea S.	Buckley class	Frigate	1958–67	6	Converted from Buckley class in 1943–44
Korea S.	Cannon class	Frigate	1955–56	2	Built 1944; re-sold to Philippines in 1977
Korea S.	Diver class	Support ship	1977–79	2	
Korea S.	Fletcher class	Destroyer	1962–68	3	Purchased in 1977
Korea S.	FS-330 type	Support ship	1950–56	5	
Korea S.	Gearing class	Destroyer	1970–72	2	Loaned 1972; purchased 1977

Country	Class	Type	Years	Number	Notes
Korea S.	LCU-501 class	Landing craft	1959–60	1	Prior to licensed production of 3; built in 1943
Korea S.	LSM type	Landing craft	1954–56	12	Loaned 1955–56 and purchased 1974
Korea S.	LSMR type	Landing ship	1959–60	1	
Korea S.	LSSL type	Support ship	1952–53	5	Transferred in the 1950s; deleted 1960–62
Korea S.	LST 1-510	Landing ship	1950–56	11	
Korea S.	LST 511-1152	Landing ship	1951–59	12	
Korea S.	PC-452 type	Patrol craft	1949–64	9	
Korea S.	PCE-827 class	Corvette	1954–61	9	
Korea S.	PT type	FAC	1951–52	4	Built 1945
Korea S.	Rudderow class	Frigate	1962–63	1	Loaned 1963 and purchased 1974
Korea S.	Sumner class	Destroyer	1971–73	2	
Korea S.	YMS class	MSC	1951–52	7	
Korea S.	YMS class	MSC	1948–56	21	
Korea S.	YO type	Support ship	1954–55	1	
Korea S.	YO type	Support ship	1945–46	1	
Korea S.	FS-330 type	Support ship	1970–71	1	
Korea S.	YO type	Support ship	1970–71	1	
Malaysia	LST 511-1152	Landing ship	1970–71	1	Loaned 1971; purchased 1974
Malaysia	LST 511-1152	Landing ship	1975–76	2	
Mexico	Gearing class	Destroyer	1981–82	2	In addition to 1 in service
Mexico	Buckley class	Frigate	1970–71	2	
Mexico	Buckley class	Frigate	1962–63	4	
Mexico	LST 511-1152	Landing ship	1971–72	2	
Mexico	Admirable class	MSO	1961–62	15	Sold at scrap value: $362 000 each
Mexico	Admirable class	MSO	1972–73	1	
Mexico	Auk class	Minelayer	1972–73	19	Minesweeping gear removed
Mexico	Edsall class	Frigate	1972–73	1	

Seller/ Buyer	Designation	Description	Years of delivery	No. deliv.	Comments
Mexico	Fletcher class	Destroyer	1969–70	2	
Netherlands	Adjutant class	MSC	1953–54	14	14 transferred under MDAP; 3 returned 1972
Netherlands	Aggressive class	MSO	1954–55	6	6 transferred under MDAP
Norway	Gardiner Bay	Training ship	1957–58	1	Used for training
Norway	LSM type	Landing craft	1952–52	2	Transferred to Turkey 1960
Norway	Auk class	Minelayer	1959–60	4	
Norway	PT type	FAC	1951	10	
Norway	LCU 1466 class	Landing craft	1958	1	
Pakistan	Tench class	Submarine	1963–64	1	Sunk in 1971 war with India
Pakistan	Bluebird class	MSC	1955–63	8	Also designated Adjutant class
Pakistan	Gearing class	Destroyer	1976–77	2	
Pakistan	Gearing class	Destroyer	1980	2	In addition to 2 delivered 1977
Pakistan	Gearing class	Destroyer	1982–83	2	
Pakistan	Brooke class	Frigate	1988	4	Mix of Brooke and Garcia class frigates and 1 repair ship leased for $6.3 m annually
Pakistan	Garcia class	Frigate	1988	4	
Panama	LSMR type	Landing ship	1975	1	
Peru	LSM type	Landing craft	1959	2	
Peru	LST 1-510	Landing ship	1949–50	4	Delivered via UK; all scrapped by 1950s
Peru	YMS class	MSC	1947–50	4	Deleted 1974
Peru	LST 511-1152	Landing ship	1977	1	Purchased from civilian firm
Peru	LST 1-510	Landing ship	1947	1	
Peru	PGM-71 class	Patrol craft	1966	1	Prior to licence production of 1 ship
Peru	Guppy-2 class	Submarine	1974–75	2	
Peru	Abtao class	Submarine	1952–57	4	Refitted 1965–1967
Peru	Tacoma class	Frigate	1948	1	Purchased 1948, modernized 1952

Country	Class	Type	Year	No.	Notes
Peru	Bostwick class	Frigate	1951	3	MDAP; torpedo tubes removed
Peru	Fletcher class	Destroyer	1960–61	2	Both ships received helicopter landing pads in 1975
Peru	Parish class	Landing ship	1983–84	4	
Philippinnes	Buckley class	Frigate	1960–61	1	Built 1943
Philippinnes	Edsall class	Frigate	1975	1	Acquired from S. Vietnam
Philippinnes	Barnegat class	Support ship	1975	4	Acquired from S. Vietnam
Philippinnes	Admirable class	MSO	1975	1	Acquired from S. Vietnam; minesweeping gear removed
Philippinnes	Admirable class	MSO	1947–48	2	Minesweeping gear removed upon transfer
Philippinnes	Auk class	Minelayer	1964–67	2	Minesweeping gear replaced with torpedoes
Philippinnes	Aggressive class	MSO	1971–72	2	Built 1953–54
Philippinnes	PCE-827 class	Corvette	1947–48	5	ASW-armed
Philippinnes	PCE-827 class	Corvette	1975–76	3	
Philippinnes	PC–452 type	Patrol craft	1946–48	6	
Philippinnes	PC–452 type	Patrol craft	1957–58	1	
Philippinnes	PC–452 type	Patrol craft	1967–68	1	
Philippinnes	PC–452 type	Patrol craft	1975	1	
Philippinnes	PGM-71 class	Patrol craft	1975	1	Acquired from Cambodia
Philippinnes	LST 1-510	Landing ship	1971–78	6	Acquired from S. Vietnam
Philippinnes	LST 511-1152	Landing ship	1968–76	18	
Philippinnes	LSM type	Landing craft	1959–62	3	
Philippinnes	LSM type	Landing craft	1975	2	Acquired from S. Vietnam
Philippinnes	LSIL type	Landing craft	1975	4	Acquired from S. Vietnam
Philippines	Cannon class	Frigate	1977–80	3	
Philippines	LCU 1466 class	Landing craft	1975	3	
Philippinnes	Bostwick class	Frigate	1966–67	1	Refurbished in S. Korea before delivery

Seller/ Buyer	Designation	Description	Years of delivery	No. deliv.	Comments
Portugal	Kellar class	Support ship	1972	1	
Portugal	J C Butler class	Destroyer	1957	2	
Saudi Arabia	LCU 1610 class	Landing craft	1975–76	4	
Singapore	Bluebird class	MSC	1974–75	2	
Singapore	LST 511-1152	Landing ship	1970–75	6	
Spain	Fletcher class	Destroyer	1957–61	5	Leased in 1957, all purchased 1972
Spain	Balao class	Submarine	1959–74	5	
Spain	Parish class	Landing ship	1971–72	3	Built 1952–53; leased 1971–72, bought 1976
Spain	Guppy-2 class	Submarine	1971–74	2	
Spain	Gearing class	Destroyer	1972–73	5	Built 1944–45 transferred under MDAP and bought 1978; no ASROC installed
Spain	Independence	Aircraft-carrier	1968	1	*Dedalo*
Spain	Parish class	Landing ship	1976	3	Purchased in 1978
Spain	LSD-13 class	AALS	1971	1	Purchased in 1974; fitted with helicopter platform
Spain	LSM type	Landing craft	1960	3	Medium landing ship transferred Mar. 1960
Spain	Adjutant class	MSC	1954–59	8	
Spain	Aggressive class	MSO	1953–56	4	All 4 ships purchased 1974
Spain	Paul Revere class	Landing ship	1980	2	
Spain	LCU 1466 class	Landing craft	1972	2	Purchased in 1976
Spain	Bostwick class	Frigate	1949–51	4	
Taiwan	Gearing class	Destroyer	1980	1	Two more purchased for spares 1980–81
Taiwan	Asheville class	Patrol craft	1984	3	Reportedly transferred 1984 from the Massachusetts Maritime Academy
Taiwan	Amphion class	Support ship	1973–74	1	
Taiwan	Auk class	Minelayer	1963–68	4	Built 1942–45; used as corvettes
Taiwan	Benson class	Frigate	1953–54	2	Built in 1940; deleted in 1975

	Class	Type	Years	No.	Notes
Taiwan	Benson class	Frigate	1954–59	2	Built 1940–42; on loan; deleted 1975–76
Taiwan	Buckley class	Frigate	1959–67	11	Built 1943–44 and converted to high speed transports during WWII
Taiwan	C1-M-AV1 type	Support ship	1971–72	1	Former merchant ship
Taiwan	Diver class	Support ship	1976–77	1	
Taiwan	Fletcher class	Destroyer	1966–71	4	
Taiwan	Guppy-2 class	Submarine	1972–73	2	
Taiwan	Liberty type	Support ship	1971–72	1	
Taiwan	LCU-501 class	Landing craft	1957–59	9	In addition to 6 abandoned before 1949; built during WWII
Taiwan	LCU-501 class	Landing craft	1963–64	7	Built during WWII; delivered 1960 and purchased 1976
Taiwan	LSD-1 class	AALS	1959–60	1	Ex-US Navy
Taiwan	LSD-13 class	AALS	1973–74	1	3 more received for cannibalization
Taiwan	LSIL type	Landing craft	1953–54	6	In addition to 8 evacuated from mainland China in 1949; refurbished during 1960s
Taiwan	LSM type	Landing craft	1958–62	7	2 more received for spares
Taiwan	LSSL type	Support ship	1953–54	3	Built during WWII
Taiwan	LST 1-510	Landing ship	1956–60	3	In addition to 8 evacuated from mainland China in 1949; incl 5 from merchant service in 1955; rebuilt in the 1960s
Taiwan	LST 511-1152	Landing ship	1954–61	20	
Taiwan	Mark class	Support ship	1970–71	1	Gun-armed cargo ship
Taiwan	Patapsco class	Support ship	1960–72	3	Gun-armed tankers
Taiwan	PC-452 type	Patrol craft	1953–59	15	Built in 1943; most deleted in early 1970s; 3 transferred to Customs service
Taiwan	Rudderow class	Frigate	1967–68	1	Built 1949
Taiwan	Sumner class	Destroyer	1968–74	8	3 armed with Gabriel ShShMs

Seller/ Buyer	Designation	Description	Years of delivery	No. deliv.	Comments
Taiwan	YO type	Support ship	1949	1	
Taiwan	Gearing class	Destroyer	1972–73	3	Bought by Taiwan after Spain rejected them; all configured for ASW
Taiwan	Gearing class	Destroyer	1972	1	ASROC launchers removed
Taiwan	Gearing class	Destroyer	1970–71	1	Helicopter deck removed
Taiwan	Gearing class	Destroyer	1976–78	4	1 more transferred for spares in 1977; armed with Gabriel ShShMs after sale
Thailand	SC type	Patrol craft	1957–59	3	
Thailand	PC-452 type	Patrol craft	1951–52	4	In addition to 4 received 1947–49
Thailand	LSSL type	Support ship	1965–66	1	
Thailand	LCI type	Landing craft	1949–50	2	
Thailand	LSM type	Landing craft	1961–62	1	In addition to 2 received in 1946
Thailand	LST 511-1152	Landing ship	1961–75	4	
Thailand	LST 1-510	Landing ship	1953–54	1	
Thailand	Cannon class	Frigate	1958–59	1	Transferred under MDAP in 1959 and sold 1975; built 1943–44
Thailand	Asheville class	Frigate	1950–51	2	
Tunisia	Edsall class	Frigate	1973	1	Built 1943–44
Turkey	LSM type	Landing craft	1952	3	
Turkey	Balao class	Submarine	1948–60	11	
Turkey	LSM type	Landing craft	1966–67	4	
Turkey	Fletcher class	Destroyer	1967–69	5	
Turkey	Adjutant class	MSC	1970	3	Supplied from Belgium and France
Turkey	Sumner class	Destroyer	1970–72	2	
Turkey	Balao class	Submarine	1970–73	10	1 upgraded to Guppy IA standard; 7 upgraded to Guppy IIA standard; 3 upgraded to Guppy III standard

Country	Class	Type	Years	No.	Notes
Turkey	Gearing class	Destroyer	1970–80	9	
Turkey	Guppy-2 class	Submarine	1977–79	2	
Turkey	Tang class	Submarine	1979–80	1	Originally ordered by Iran
Turkey	Gearing class	Destroyer	1980–81	1	ASW version (Carpenter class)
Turkey	Gearing class	Destroyer	1981–83	2	In addition to 10 in service; 1 bought, 1 leased
Turkey	Dixie class	Support ship	1982–83	1	On lease; to support Gearing class
Turkey	Tang class	Submarine	1983	1	Designation unconfirmed
Turkey	Parish class	Landing ship	1973–74	2	
Turkey	Asheville class	Frigate	1972–73	2	
Uruguay	Dealey class	Frigate	1972	1	
Uruguay	Gearing class	Destroyer	1983	1	
Uruguay	LCI type	Landing craft	1972	2	
Uruguay	Bostwick class	Frigate	1951	2	Guns and torpedoes removed
Uruguay	Cannon class	Frigate	1952	2	Modernized late 1960s, all guns removed
Uruguay	PC-452 type	Patrol craft	1964	1	Deleted 1969
Uruguay	Auk class	Minelayer	1966	1	Loaned 1966; purchased Aug 1976
Venezuela	Parish class	Landing ship	1973	1	
Venezuela	Sumner class	Destroyer	1973–74	2	
Venezuela	Balao class	Submarine	1960	1	Overhauled 1962; for training
Venezuela	LSM type	Landing craft	1959–60	4	2 deleted 1979
Venezuela	LST-1173 class	Landing ship	1973–77	1	Loaned 1973, purchased 1977
Venezuela	Guppy-2 class	Submarine	1973	1	Refitted in Argentina
Vietnam (S.)	Admirable class	MSO	1961–70	4	Built in 1943–44
Vietnam (S.)	Barnegat class	Support ship	1970–72	7	
Vietnam (S.)	Edsall class	Frigate	1971	2	
Vietnam (S.)	PC-452 type	Patrol craft	1959–60	1	
Vietnam (S.)	PCE-827 class	Corvette	1960–70	3	
Vietnam (S.)	LCU 1466 class	Landing craft	1957–71	15	

Seller/ Buyer	Designation	Description	Years of delivery	No. deliv.	Comments
Vietnam (S.)	LSM type	Landing craft	1960-65	3	
Vietnam (S.)	LSSL type	Support ship	1964-66	5	First sold to Japan in 1953-64
Vietnam (S.)	LST 1-510	Landing ship	1969-70	1	
Vietnam (S.)	LST 511-1152	Landing ship	1961-70	5	Built 1943-45
USSR					
Albania	Whiskey class	Submarine	1958-60	2	
Algeria	Polnocny class	Landing ship	1976	1	
Algeria	Romeo class	Submarine	1982-83	2	Not returned to USSR as planned but retained; terms of transfer unclear
Algeria	T-43 class	MSO	1968	2	
Algeria	P-6 class	FAC	1963-68	12	
Algeria	Komar class	FAC	1966	6	
Algeria	Osa-1 class	FAC	1967	3	
Algeria	SO-1 class	Patrol craft	1965-67	6	
Algeria	Osa-2 class	FAC	1975-81	9	
Angola	Shershen class	FAC	1976-83	5	
Angola	Polnocny class	Landing ship	1977-79	3	
Angola	Osa-2 class	FAC	1981-83	6	
Angola	Yevgenia class	MSC	1985-87	2	
Bulgaria	Osa-2 class	FAC	1977-84	4	
Bulgaria	Sonya class	MSC	1981-83	2	
Bulgaria	Vydra class	Landing craft	1969-80	20	Fitted with Spin Trough Radar
Bulgaria	P-2 class	FAC	1957-60	24	Possibly assembled; half deleted 1976, last 4 possibly deleted by 1982
Bulgaria	Yevgenia class	MSC	1970-78	4	
Bulgaria	Kronstadt class	Patrol craft	1957	2	
Bulgaria	SO-1 class	Patrol craft	1963	6	Unconfirmed

Country	Class	Type	Year	Number	Notes
Bulgaria	T-43 class	MSO	1953	3	One cannibalized
Bulgaria	T-301 class	MSC	1955	4	
Bulgaria	MV class	Submarine	1950–52	3	3 transferred early 1950s; 2 exchanged for 2 Whiskey class 1958
Bulgaria	Romeo class	Submarine	1972–73	2	Deleted by 1982
Bulgaria	Whiskey class	Submarine	1958	2	Deleted late-1970
Bulgaria	Riga class	Frigate	1957–58	2	Transferred 1957–58; refitted 1980–81
Bulgaria	Osa-1 class	FAC	1969–71	3	
Bulgaria	Vanya class	MSC	1983–85	2	In addition to 4 delivered 1970–71
Bulgaria	Romeo class	Submarine	1984–85	1	
Bulgaria	Romeo class	Submarine	1985–86	1	Unconfirmed
Cape Verde	Shershen class	FAC	1979	2	1 without torpedo tubes
China	Artillerist class	Patrol craft	1954–55	6	
China	S class	Submarine	1954–55	8	
China	T-43 class	MSO	1954–55	4	2 returned in 1960
China	SO-1 class	Patrol craft	1960	2	
China	P-4 class	FAC	1951–53	70	Some possibly assembled in China
China	Komar class	FAC	1958–60	7	
China	Shershen class	FAC	1979	2	
Congo	Osa-1 class	FAC	1972–74	6	
Cuba	Turya class	Hydrofoil FAC	1978–81	6	
Cuba	Whiskey class	Submarine	1978–79	1	
Cuba	Sonya class	MSC	1979–80	2	
Cuba	Osa-2 class	FAC	1976–82	13	
Cuba	Yevgenia class	MSC	1976–82	10	
Cuba	Polnocny class	Landing ship	1981–82	2	
Cuba	Sonya class	MSC	1984–85	2	
Cuba	Yevgenia class	MSC	1984	2	

Seller/ Buyer	Designation	Description	Years of delivery	No. deliv.	Comments
Cuba	P-4 class	FAC	1961–64	12	
Cuba	P-6 class	FAC	1961–62	16	
Cuba	Komar class	FAC	1961–66	18	
Cuba	Kronstadt class	Patrol craft	1961–62	6	
Cuba	SO-1 class	Patrol craft	1963–67	12	
Cuba	Foxtrot class	Submarine	1978–80	2	
Cuba	Foxtrot class	Submarine	1982–84	1	
Cuba	Stenka class	FAC	1984–85	3	Torpedo tubes removed before transfer
Egypt	SMB-1 class	Landing craft	1964–65	4	
Egypt	Vydra class	Landing craft	1967–69	10	
Egypt	Polnocny class	Landing ship	1972–74	3	
Egypt	T-43 class	MSO	1955–71	7	
Egypt	T-301 class	MSC	1961–62	2	
Egypt	Yurka class	MSO	1969–71	4	
Egypt	SO-1 class	Patrol craft	1961–71	12	
Egypt	P-6 class	FAC	1955–70	36	Some built in Alexandria
Egypt	Shershen class	FAC	1966–68	7	
Egypt	Komar class	FAC	1961–67	8	
Egypt	Osa-1 class	FAC	1965–66	12	
Egypt	MV class	Submarine	1956–57	1	
Egypt	Skory class	Destroyer	1955–56	2	Exchanged for similar ships in 1967; re-equipped with Styx ShShMs taken from Komar class FACs
Egypt	Skory class	Destroyer	1961–62	2	
Egypt	Romeo class	Submarine	1966–69	6	
Egypt	Whiskey class	Submarine	1971–72	2	
Egypt	Whiskey class	Submarine	1962	1	Replaced 2 Whiskey class returned to USSR

Country	Class	Type	Years	Number	Notes
Egypt	Whiskey class	Submarine	1957–58	7	
Eq. Guinea	P-6 class	FAC	1977–78	1	
Ethiopia	Osa-2 class	FAC	1978–81	4	
Ethiopia	Polnocny class	Landing ship	1981	1	In addition to 1 delivered 1981
Ethiopia	Polnocny class	Landing ship	1982–83	1	
Ethiopia	Petya-2 class	Frigate	1982–83	1	
Ethiopia	Petya-2 class	Frigate	1984	1	In addition to 1 delivered 1983
Ethiopia	Barnegat class	Support ship	1962	1	Loaned in 1962 bought in 1976
Ethiopia	Mol class	FAC	1978	2	
Finland	Riga class	Frigate	1963–64	2	1 deleted, 1 converted to minelayer 1979
Gui. Bisseau	P-6 class	FAC	1975–76	4	
Gui. Bisseau	Shershen class	FAC	1979	2	
GDR	Hvidbjornen	Support ship	1950	1	Built in Denmark 1928; seized by USSR 1945; scrapped in 1968
GDR	R-Boat type	MCM	1950	6	
GDR	Riga class	Frigate	1955–59	4	
GDR	P-6 class	FAC	1956–59	27	
GDR	Shershen class	FAC	1966–76	7	
GDR	Tarantul class	Corvette	1982–86	5	
Guinea	T-58 class	MSO	1978–79	1	
Guinea	P-6 class	FAC	1965–72	6	
Guinea	Shershen class	FAC	1978–79	4	
India	Osa-2 class	FAC	1975–77	8	
India	Osa-1 class	FAC	1970–71	8	
India	Ugra class	Support ship	1967–68	1	Converted to submarine rescue ship
India	T-58 class	MSO	1970–71	1	
India	Charlie-1 class	SSN	1985–88	1	Nuclear-powered submarine leased for 3 years

Seller / Buyer	Designation	Description	Years of delivery	No. deliv.	Comments
Indonesia	Kronstadt class	Patrol craft	1957–58	14	All scrapped 1973
Indonesia	Skory class	Destroyer	1961–1964	3	Last 2 scrapped 1981–83
Indonesia	Whiskey class	Submarine	1958–62	12	All deleted by 1981
Indonesia	Komar class	FAC	1960–65	12	
Indonesia	Riga class	Frigate	1962–64	7	
Indonesia	P-6 class	FAC	1960–62	14	Last 10 probably only for spares
Indonesia	BK class	Patrol craft	1961–62	18	Reportedly transferred 1962
Indonesia	Bunju type	Tanker	1958–59	2	Gun-armed tankers
Indonesia	T-43 class	MSO	1961–64	6	
Indonesia	Sverdlov class	Cruiser	1962	1	Deleted 1972
Indonesia	T-301 class	MSC	1962	1	
Indonesia	Don class	Support ship	1961–62	1	Unarmed submarine tender
Indonesia	Atrek class	Support ship	1961–62	1	
Iraq	Polnocny class	Landing ship	1976–79	4	One sunk by Iranian Harpoon missile 1980
Iraq	T-43 class	MSO	1968–69	2	
Iraq	P-6 class	FAC	1958–61	12	
Iraq	SO-1 class	Patrol craft	1961–62	3	
Iraq	Osa-1 class	FAC	1971–74	6	
Iraq	Osa-2 class	FAC	1973–77	8	At least 4 sunk by 1983
Kampuchea	Turya class	Hydrofoil FAC	1984–85	2	
Kampuchea	Stenka class	FAC	1984–85	1	Torpedo tubes and sonar removed
Korea N.	P-4 class	FAC	1958–60	12	Unconfirmed
Korea N.	Tral class	MSO/PC	1954–55	8	Number unconfirmed
Korea N.	Komar class	FAC	1970–72	6	
Korea N.	Osa-1 class	FAC	1970–73	8	
Korea N.	Whiskey class	Submarine	1960–61	4	Delivery schedule unconfirmed
Korea N.	Shershen class	FAC	1973–74	4	Torpedo-armed

Country	Class	Type	Years	No.	Notes
Korea N.	T-43 class	MSO	1962–63	2	
Libya	Polnocny class	Landing ship	1977–79	4	
Libya	Natya class	MSO	1980–81	2	
Libya	Natya class	MSO	1982–84	4	In addition to 2 delivered 1981
Libya	Nanuchka class	Corvette	1986	2	
Mozambique	SO-1 class	Patrol craft	1984–85	2	
Poland	Skory class	Destroyer	1957–58	2	Scrapped 1975
Poland	Kotlin class	Destroyer	1969–70	1	
Poland	MV class	Submarine	1953–55	6	Gun removed 1958–59; all scrapped by 1970
Poland	Whiskey class	Submarine	1961–69	4	
Poland	P-6 class	FAC	1955–58	20	12 scrapped 1973–75; all scrapped by 1982
Poland	Osa-1 class	FAC	1963–65	13	All in service by 1984
Poland	Kronstadt class	Patrol craft	1954–57	8	Scrapped 1973–74
Poland	Kashin class	Destroyer	1985–87	1	
Poland	Foxtrot class	Submarine	1987	1	Leased to Poland after delays in transfers of Kilo class submarines
Romania	T-301 class	MSC	1956–60	24	Approx 10 in service by 1985
Romania	M-1940 type	MSC	1955–57	4	German WW II MSCs refitted in USSR
Romania	P-4 class	FAC	1960–62	12	All scrapped by late 1970s
Romania	Osa-1 class	FAC	1961–64	6	
Romania	Kronstadt class	Patrol craft	1955–56	3	
Romania	Poti class	Corvette	1969–70	3	
Romania	Kilo class	Submarine	1986	1	Further deliveries possible
Seychelles	Turya class	Hydrofoil FAC	1986	1	Torpedo tubes removed prior to delivery
Seychelles	Turya class	Hydrofoil FAC	1980	1	Hydroplanes removed before transfer
Somalia	Osa-2 class	FAC	1975	2	
Somalia	Mol class	FAC	1976–77	4	
Somalia	P-6 class	FAC	1972	7	

Seller/Buyer	Designation	Description	Years of delivery	No. deliv.	Comments
Sri Lanka	Mol class	FAC	1975–77	1	
Syria	Osa-2 class	FAC	1977–84	10	
Syria	Osa-2 class	FAC	1985	2	
Syria	Yevgenia class	MSC	1977–81	2	
Syria	Vanya class	MSC	1972–73	2	
Syria	T-43 class	MSO	1959–62	2	
Syria	Natya class	MSO	1984–85	1	
Syria	P-4 class	FAC	1956–74	17	Replacing 1 lost in October War 1973
Syria	Komar class	FAC	1962–66	6	
Syria	Osa-1 class	FAC	1969–73	9	
Syria	Petya-2 class	Frigate	1974–75	2	Replacing war losses
Syria	Komar class	FAC	1973–74	3	Used as battery chargers for other submarines
Syria	Romeo class	Submarine	1984–85	2	
Syria	Sonya class	MSC	1985–86	1	
Vietnam	Petya-2 class	Frigate	1978	2	Gift
Vietnam	SO-1 class	Patrol craft	1979–83	8	
Vietnam	Shershen class	FAC	1979–83	14	In addition to 2 delivered 1973
Vietnam	Osa-2 class	FAC	1979–81	8	
Vietnam	Yurka class	MSO	1979	1	
Vietnam	Polnocny class	Landing ship	1979–80	3	
Vietnam	Petya-2 class	Frigate	1983–84	3	In addition to 2 in service
Vietnam	Turya class	Hydrofoil FAC	1983–86	5	
Vietnam (N.)	Komar class	FAC	1971–72	4	
Vietnam (N.)	P-4 class	FAC	1960–64	12	
Vietnam (N.)	P-6 class	FAC	1966–72	6	In addition to 6 received from China
Vietnam (N.)	Shershen class	FAC	1972–73	2	

Vietnam (N.)	SO-1 class	Patrol craft	1960–65	4	
Yemen (N.)	Osa-2 class	FAC	1981–82	2	
Yemen (N.)	Yevgenia class	MSC	1982	2	
Yemen (N.)	Ondatra class	Landing craft	1982–83	2	
Yemen (N.)	P-4 class	FAC	1968–69	4	
Yemen (S.)	Osa-2 class	FAC	1979–83	8	
Yemen (S.)	Ropucha class	Landing ship	1978–79	1	
Yemen (S.)	T-58 class	MSO	1977–78	1	Minesweeping gear removed
Yemen (S.)	SO-1 class	Patrol craft	1971–72	2	
Yemen (S.)	Mol class	FAC	1977–78	2	
Yemen (S.)	P-6 class	FAC	1972–73	2	
Yemen (S.)	Polnocny class	Landing ship	1976–77	1	
Yemen (S.)	Polnocny class	Landing ship	1972–73	2	
Yugoslavia	Osa-1 class	FAC	1964–69	10	
Yugoslavia	Shershen class	FAC	1964–68	4	Followed by licensed production of 11 ships
Yugoslavia					
Bangladesh	Kraljevica class	Patrol craft	1974–75	2	
Egypt	Type 108 FAC	FAC	1955–56	6	
Ethiopia	Kraljevica class	Patrol craft	1974–75	1	
Indonesia	Kraljevica class	Patrol craft	1958	6	
Indonesia	LCT type	Landing craft	1957–58	4	
Kampuchea	Type 108 FAC	FAC	1965	2	Gift
Sudan	Kraljevica class	Patrol craft	1969	4	
Sudan	DTM-211 class	Landing craft	1969	1	

Appendix 2. Maritime patrol aircraft and their surveillance capabilities in selected countries

Country/System	Seller	No./ordered	Role	Search capability (1000s km per hr/day)	Radar type
Argentina: marine jurisdictional claim 339 500 km²; 200-nautical-mile fishing zone					
SA 319B Alouette III	France	8	ASW	7.6/38	Omera ORB-31S1SA 330
SA-330L Puma	France	3	SAR	16/66	RDR 1400/RCA Primus 500
Bell 212 ASW	USA	10	ASW	10/51	Ferranti Seaspray
ASH-3D/H Sea King	Italy	5	ASW	49/172	SMA/APS-707
CH-47C Chinook	USA	2	SAR		Human eye
S-2E Tracker	USA	6	MR/ASW	126/440	Retractable search radar
L-188E Electra	USA	3	MR/ASW	167/668	..
SP-2H Neptune	USA	6	MR/ASW	178/712	..
Lynx Mk 23	UK	1	ASW	24/83	Ferranti Seaspray
Hughes MD500	USA	6	SAR	..	..
Australia: marine jurisdictional claim 1 854 000 km²; 12-nautical-mile fishing zone					
Searchmaster B	..	..	MS	41/162	Bendix RDR 1400
Searchmaster L	..	..	MS	81/284	Litton APS-504(v)2
P-3C/Orion update	USA	12/9	MS	77/383	AN/APS-94F SLR; AN/APS-115
SH-60B-2 Seahawk	USA	16	ASW	39/136	MEL 'Supersearcher'
Sea King HAS 50/50A	UK	6/2	ASW	47/172	AW 391
Wessex HAS 31B	UK	10	SAR	..	Special search equipment
Brazil: marine jurisdictional claim 924 000 km²; 200-nautical-mile fishing zone					
ASH-3H Sea King	USA	10/2	ASW	47/163	SMA/APS-707
EMB-110 Bandeirante	..	14	MR/SAR	..	..
EMB-111	..	12	MR	177/618	AIL AN/APS-128 SPAR-1

S-2A Tracker	USA	7	ASW	126/440	Retractable search radar
S-2E Tracker	USA	8	ASW	126/440	Retractable search radar
RC-130E Hercules	USA	5	MR/SAR	315/1.1	AN/APQ-122(V) on A
Lynx Mk 21	UK	9	ASW	24/83	Ferranti Seaspray
Westland Wasp	UK	7	SAR	:	:

Canada: marine jurisdictional claim 1 370 000 km²; 200-nautical-mile fishing zone

CS2F-3 Tracker	USA (I)	29	CS	126/440	Retractable search radar
Aurora	USA	18	MS	231/963	AN/APS-116
CHSS-2 Sea King	USA	32	ASW	:	AN/APS-503

Chile: marine jurisdictional claim 667 300 km²; 200-nautical-mile fishing zone

Falcon 200 Gardian	France	2	MR	:	Thomson-CSF Varan
SA 319B Alouette III	France	10	ASW/SAR	7.6/38	Omera ORB-31S1
206A Jetranger	USA	3	ASW/SAR	17/85	:
Embraer EMB-111	Brazil	6	MR	177/618	AIL AN/APS-128 SPAR-1

China, People's Republic: marine jurisdictional claim 281 000 km²

SA 321G Super Frelon	France	12	ASW	33/131	Omera ORB-31D/32
Be-6 Madge	:	10	MR	196/686	:
Y-8 (An-12BP)	:	1	MR	168/672	:
SH-5 (Il-28 Beagle)	:	8	Recce	65/262	:
Z-5 (Mi-4 Hound)	:	40	ASW/SAR	8/40	:
Z-9 SA 365N Dauphin 2	France (I)	10/15	ASW	9/54	Agrion 15
H-6 (Tu-16 Badger)	:	50	MR	192/672	:

Egypt: marine jurisdictional claim 50 600 km²

E-2C Hawkeye	USA	5	AEW&C	77/310	AN/APS-138/125
Il-28 Beagle	USSR	13	MR	65/262	:
TU-16 Badger	USSR	12	MR	192/672	:
Sea King Mk 47	UK	5	ASW	49/172	AW 391

Country/System	Seller	No./ordered	Role	Search capability (1000s km per hr/day)	Radar type
India: marine jurisdictional claim 587 500 km²; 200-nautical-mile fishing zone, 200-nautical-mile EEZ					
SA 319B Chetak	France (1)	21	ASW/SAR	7.6/38	Omera ORB-31S1
Br 1050 Alize	France	8	ASW	25/150	Thomson-CSF Iguane
Do-228	Germany(1)	60	MS	121/424	MEL Marec 2
Il-38 May	USSR	3	MR	215/859	..
Ka-25 Hormone	USSR	5	ASW	13/65	..
Ka-27 Helix	USSR	18	ASW	16/80	..
BN-2 Maritime Defender	UK	18	MR	83/333	Bendix RDR 1400
Tu-142 Bear-F	USSR	5	MR	502/1.76	..
Sea King Mk 42/42A/B[c]	UK	35	ASW	49/172	AW 391
Indonesia: marine jurisdictional claim 1 577 000 km²; 200-nautical-mile fishing zone, 200-nautical-mile EEZ					
B-737-200 Surveiller	USA	3	MR	188/656	SLAM/MR
Searchmaster B	Australia	12	MR	41/162	Bendix RDR 1400
Searchmaster L	Australia	6	MR	81/284	Litton APS-504(V)2
HU-16B Albatross	USA	5	SAR	..	Nose radome
C-130H-MP Hercules	USA	1	MR	315/1.1	SSR,SLAR AN/APQ-122?
NAS-332 Super Puma	France	26	ASW	38/133	Omera ORB 3214
Wasp (HAS Mk 1)	UK	14	ASW	..	..
Iran[a]: marine jurisdictional claim 45 400 km²; 50-nautical-mile fishing zone					
AB-204 ASW	Italy	12	ASW	10/51	Ferranti Seaspray
ASH-3D Sea King	Italy	12	ASW	19/105	SMA/APS-707
P-3F Orion	USA	2	MR/ASW	77/383	AN/APS-94F SLR

C-130H-MP Hercules	USA	5	MR	315/1.1	AN/APS-115 SSR,SLAR AN/APQ-122

Iraq: marine jurisdictional claim 200 km²

SA 321 H Super Frelon	France	8	Attack	33/131	Omera ORB-31D/32
AB-212 ASW	Italy	8	ASW	10/51	Ferranti Seaspray

Israel: marine jurisdictional claim 6800 km²

206A Jetranger	USA	2	SAR	::	..
Bell 212	USA	25	SAR/ECM	10/51	..
E-2C Hawkeye	USA	4	AEW	77/310	AN/APS-138/125
1124 Sea Scan	USA	7	CP	278/973	Litton APS-504(V)2

Japan: marine jurisdictional claim 112 600 km²; 200-nautical-mile fishing zone

Maritime Patrol 200T		15	MR	::	2 alt. search radars DITACS
Bell-212	USA	29	CP	10/51	..
Kawasaki P-2J	USA (I)	78	ASW/MR	267/980	AN/APS-80J
P-3C Update II	USA (I)	24	ASW/MR	77/383	AN/APS-94F SLR AN/APS-115
P-3C Update III	USA (I)	50	MR	300/997	Japanese avionics suite
Mitsubishi SH-3A	USA (I)	64	ASW	47/163	SMA/APS-707
S-61/S-62	USA (I)	12	SAR	26/91	SMA/APS-707
PS-1/US-1A	..	22	ASW/SAR	..	Litton AN/APS-504

Korea, South: marine jurisdictional claim 101 600 km²; 200-nautical-mile fishing zone

S-2A/F Tracker	USA	20	ASW	126/440	Retractable search radar
500MD/ASW Defender	USA	10	ASW	::	Search radar

Country/ System	Seller	No./ ordered	Role	Search capability (1000s km per hr/day)	Radar type
Kuwait: marine jurisdictional claim 4100 km²					
AS 332F Super Puma	France	12	ASV	38/133	Omera ORB 3214
SA 365F Dauphin 2	France	4	SAR	9/54	Agrion 15
Libya: marine jurisdictional claim 98 600 km²					
SA 321 M Super Frelon	France	8	SAR	33/131	Omera ORB-31D/32
Mi-14 Haze	USSR	12	ASW	16/80	..
Alouette III	France	12	..	Human eye	..
Malaysia: marine jurisdictional claim 138 700 km²; 200-nautical-mile fishing zone, 200-nautical-mile EEZ					
C-130H-MP Hercules	USA	3	MR/SAR	315 *1.1*	AN/APQ-122
Morocco: marine jurisdictional claim 81 000 km²; 200-nautical-mile fishing zone, 200-nautical-mile EEZ					
HH-43B Huskie	USA	4	SAR	..	..
New Zealand: marine jurisdictional claim 1 058 000 km²; 200-nautical-mile fishing zone, 200-nautical-mile EEZ					
P-3K Orion	USA	6	MR	245/997	AN/APS-94F SLR AN/APS-115 [AN/APS-134(V)]
Westland Wasp HAS Mk 1	UK	10	ASW	..	..
F-27 Maritime	Netherlands	3	SAR	300/*1.05*	Litton AN/APS-504(V)2
Pakistan: marine jurisdictional claim 92 900 km²; 200-nautical-mile fishing zone, 200-nautical-mile EEZ					
SA 319B Alouette III	France	4	SAR	7.6/38	Omera ORB-31S1
Br 1150 Atlantic 1	France	4	ASW/MR	90/540	CSF

Sea King Mk 45	UK	6	ASW	49/172	AW 391
P-3C	USA	../3		77/383	AN/APS-94F SLR AN/APS-115
F-27	Netherlands	2	MR	300/1.05	Litton AN/APS-504

Papua New Guinea: marine jurisdictional claim 684 200 km²

N22B Missionmaster	Australia	6	MR	..	Search radar

Philippines: marine jurisdictional claim 551 400 km²; 200-nautical-mile fishing zone, 200-nautical-mile EEZ

F-27 Maritime	Netherlands	3	MR	300/1.05	Litton AN/APS-504(V)2
BO 105	Germany, FR	5	ASW/SAR	26/92	Search radar
BN-2A Islander	UK (l)	9	MR/SAR	83/333	Bendix
BN-2A Defender	UK	10	MR	83/333	Bendix

Saudi Arabia: marine jurisdictional claim 54 900 km²

SA 365F Dauphin 2	France	4	SAR	9/54	Omera ORB-32
AS-332 Super Puma	France	6	ASV	9/54	Omera ORB-32
E-3A Sentry	USA	5	AWACS	..	..

Seychelles: marine jurisdictional claim: continental shelf; 200-nautical-mile fishing zone, 200-nautical-mile EEZ

Merlin IIIB	Belgium	1	MR	..	Search radar
BN-2 Maritime Defender	UK	1	MR	83/333	Bendix RDR 1400
SA 319B Chetak	India	2	SAR	..	..

Singapore: marine jurisdictional claim 100 km²

E-2C Hawkeye	USA	4	AEW/MR	77/310	AN/APS-138/125

South Africa: marine jurisdictional claim 295 000 km²; 200-nautical-mile fishing zone

P.166S Albatross	Italy	18	CP	..	Nose radar
Westland Wasp HAS Mk 1	UK	9	ASW	..	..

Country/System	Seller	No./ordered	Role	Search capability (1000s km per hr/day)	Radar type
Taiwan: marine jurisdictional claim 114 400 km²					
HU-16B Albatross	USA	8	SAR	..	Nose radar
S-2 Tracker version A	USA	9	ASW	126/440	Retractable search radar
version E	USA	20	MR	126/440	..
version F	USA	9	ASW/MR	126/440	..
500MD/ASW	USA	12	ASW	..	Search radar
Thailand: marine jurisdictional claim 94 700 km²; 12-nautical-mile fishing zone					
Canadair CL-215	Canada	2	SAR	126/440	Search radar
F-27 Maritime	Netherlands	3	MR	300/1.05	Litton AN/APS-504(V)2
Searchmaster B	Australia	5	MR	41/162	Bendix RDR 1400
HU-16B Albatross	USA	2	SAR	..	Nose radome
S-2F Tracker	USA	10	ASW/MR	126/440	Retractable search radar
United Arab Emirates: marine jurisdictional claim 17.3 km²					
AS 332F Super Puma	France	4	ASW	38/133	Omera ORB 3214
BN-2 Maritime Defender	UK	2	MR	83/333	Bendix RDR 1400

[a] Iran also uses Swiss-supplied Pilatus PC-7 aircraft for spotting and target acquisition.

Sources:

Jane's Fighting Ships 1986-87 (Jane's: London 1986); *Jane's Fighting Ships 1987-88* (Jane's: London, 1987); *Jane's Fighting Ships 1988-89* (Jane's: Coulsden, 1988); *Combat Fleets of the World 1986/87* (Arms and Armour Press: US Naval Institute, 1986); *Jane's Weapon Systems 1987-88* (Jane's: London, 1987); *Jane's Weapon Systems 1988-89* (Jane's: Coulsden, 1988); *Air Forces of the World 1986* (Interavia: Geneva, 1986); *International Air Forces & Military Aircraft Directory* (Aviation Advisory Services Ltd: Stapleford, England); *Aviation Week & Space Technology*, last week of Sep. 1987; *Flight International*, last week of Sep. 1987.

Methods:

This is an inventory of all aircraft in service in the listed countries dedicated to the task of maritime reconnaissance or which one would normally expect to find at sea—though clearly in cases of necessity any aircraft could assist in providing information. The countries listed exclude members of the major alliances for two reasons.

First, for the superpowers the nature of their procurement is so different from that of other countries that, while they do participate in the arms trade, it is in a way very far removed from the countries here. Second, because these countries have forces dedicated to the function of maritime patrol which, while nominally under the command of the armed forces and certainly a factor in contingency planning, maintain a distinct profile in day to day operations.

The data on maritime geography list first, the sea space over which the country concerned claims some jurisdiction; and second, the specific nature of claims—whether the country has claimed an EEZ or an exclusive fishing zone, for example. The columns are mostly self-explanatory except for columns two and five, the seller and search capability.

The (l) in the seller column indicates that a system is produced under licence in the country in whose inventory it is listed. The search capability is a product of the following process: (*a*) identify the aircraft in service dedicated to the task of maritime reconnaissance or which one would normally expect to find at sea—ASW helicopters on board ships for example; (*b*) match radars to the aircraft since the same airframe can mount different radar; the same aircraft in the inventories of different countries have different capabilities; and (*c*) calculate the search capability per hour by dividing the maximum range for one flight (in km) by the maximum speed of which the aircraft is capable (in km per hour) and adding a fraction to allow for the fact that none of these missions is likely to be undertaken at full speed or to the outer limits of range. This gives an estimate of the aircraft's hourly progress, which is multiplied by the range of the radar on the aircraft. The daily figure is based on this but multiplied by the number of sorties an aircraft might reasonably be expected to make per day. In countries with limited ground support facilities and aircrew it would not be possible to turn around aircraft continuously.

Index